T0283544

Claim Your NFT!

Please visit **AmandaCassatt.com** and follow the instructions there to claim your NFT to join the Web3 Marketing community, using the 20-character code below.

Scratch off to view code.

1958

Praise for *Web3 Marketing*

"Amanda is the original storyteller of crypto, having been there since the beginning. She has since helped many companies and founders find their voice to be the authors of their own history. Her contributions to the advent of Web3 continue to be tremendous."

—**Min Teo,**
Managing Partner, Ethereal Ventures

"With the advent of Web3 technology, the rules of the marketing game are changing yet again. Cassatt, who was arguably the first marketer in the space, provides a comprehensive guide that gives a terrific overview of the history of crypto, as well as offers explicit, actionable advice on how a marketer can capitalize on the new tools that are available and on the horizon."

—**Josh Quittner,**
Founder and CEO of Decrypt

"Web3 may have been conceived by millennials but it's really being built for Gen Z; for their culture of empowerment, digital identity, and their decentralized ambitions for the future. Amanda was there when Web3 took shape and within her book explains clearly how marketers need to shape Web3 value for this future generation who are about to inherit the earth. Compelling storytelling and practical directions, she has clearly already stepped on all the land mines for us and found the path for brands and business to follow."

—**Dickon Laws,**
Global Head of Innovation, Ogilvy

Web3 Marketing

Amanda Cassatt
CEO, Serotonin

Web3 Marketing

A Handbook for the **Next** Internet Revolution

Marketing

WILEY

Library of Congress Cataloging-in-Publication Data:

Names: Cassatt, Amanda, author.
Title: Web3 marketing : a handbook for the next Internet revolution /
   Amanda Cassatt.
Description: Hoboken, New Jersey : John Wiley & Sons, Inc., [2023] |
   Includes bibliographical references and index.
Identifiers: LCCN 2022052843 (print) | LCCN 2022052844 (ebook) | ISBN
   9781394171958 (hardback) | ISBN 9781394172054 (adobe pdf) | ISBN
   9781394171965 (epub)
Subjects: LCSH: Internet marketing.
Classification: LCC HF5415.1265 .C38 2023 (print) | LCC HF5415.1265
   (ebook) | DDC 658.8/72--dc23/eng/20230124
LC record available at https://lccn.loc.gov/2022052843
LC ebook record available at https://lccn.loc.gov/2022052844

Cover Design: Wiley
Cover Image: © Chromie Squiggle #2172, Erick Calderon a/k/a Snowfro. All Rights Reserved.

SKY10042700_021323

To Sam, for your patience while I wrote this book on our vacation.

Contents

Preface

It was 2019 and my husband and I were badly delayed in the Geneva airport, when a screen caught my attention. I looked up blearily from the box of Kambly biscuits in my crumb-covered lap to confirm that it wasn't a trick of the eye. It wasn't. Prominently displayed on airport monitors was the word "Ethereum," accompanied by its price chart.

We had both started contributing to the Ethereum Project when only a small group of people were aware of it. My job as a marketer was to share Ethereum with the world, and now here it was, considered important enough by the Geneva airport's administrators to deserve a place alongside Bitcoin and the top global stocks. Every marketer dreams of the day when a project they've taken from obscurity "goes mainstream." Excited, I leapt up to snap a photo, showering the carpet with chocolatey bits. This was one of many moments, starting in 2017 and continuing to the present, that would add to my conviction that our movement—building the decentralized third generation of the web—had arrived, and that it was here to stay.

In 2016, when I began doing the work that an editor from Wiley would eventually call *Web3 marketing*, almost no one had heard of Web3. We barely used the term ourselves. I certainly had no idea that as ConsenSys's chief marketing officer, working under a mandate to bring Ethereum to market, I would become the first Web3 marketer, hire and train the first Web3 marketing team, and bring to market many of the key products underlying Web3. At the time, we were a global network of nerds with a penchant for the esoteric. The nascent industry in which we worked was called *crypto*. An ether token cost a few dollars, and nobody owned any NFTs. Considering our modest beginnings, our ambitions were laughably grand. We had full conviction that Ethereum would become not only the next great computing platform, but a new foundation for the global financial system, the substrate for building an adjacent economy governed fairly and transparently by code. The offices where we worked practically glowed with expectation.

The future we envisioned back then—one in which millions of people would safeguard their own assets in user-controlled wallets, benefit from using web applications without becoming the real product being sold, and get paid to create ownable pieces of internet property—is becoming a reality. Admittedly, the process has been slower than many of us expected. Nonetheless, crypto is in the news on a daily basis now, and the leading user-controlled wallet, MetaMask, has 21 million monthly active users. There are 28.6 million crypto wallets that hold pieces of internet property known as non-fungible tokens (NFTs).[1] Meanwhile Ethereum has grown a nearly $200 billion market capitalization, second only to Bitcoin with its near $400 billion.[2] According to estimates, over 200,000 software developers have learned Solidity, the main language for programming applications on the Ethereum blockchain.[3] And in 2021, Ethereum's $11.6 trillion transaction volume surpassed even the traditional payments giant Visa, with $10.4 trillion.[4] That's a lot of money, and a huge number of people who know about and use Ethereum, considering we started from zero.

If this sounds impressive, though, let us put it into perspective. Ethereum may have transferred more assets by value, but the 1 million transactions per day on Ethereum are no match for the Visa network, which processes over 150 million.[5] The total market cap of all cryptocurrencies, which has at times exceeded $1 trillion, pales in comparison to the total US gross domestic product (GDP) of about $25 trillion, or the global GDP of more than $96 trillion.[6] Of the total number of the estimated 31 million software developers in the world, fewer than 1% are familiar with Solidity.[7] And it's important to keep in mind that 37% of the global population hasn't used the internet, let alone collected NFTs.[8] Undeniably, our movement still has a long way to go.

We called it *crypto*—short for cryptocurrency—at the beginning, because its first applications were financial. Bitcoin was the first successful digital money system built on a decentralized blockchain. Many of the early applications on Ethereum had to do with financial value. Between 2012 and 2019, most of the value coming into the crypto ecosystem arrived through exchanges, superhighways like Coinbase where people could trade in their fiat currencies like dollars for investments in bitcoin or ether.[9] Not only the money, but the people working in the industry came from technology and finance backgrounds.

Our nascent industry's parents were finance and technology, but starting in 2020 with the rise of NFTs, that changed. The crypto industry began intersecting with the arts, entertainment, fashion, and media. This brought an entirely new wave of personalities into our space, from creators to entrepreneurs and professionals, who reshaped its character. Suddenly, the audience downloading MetaMask included buyers of digital handbags for their avatars to wear in the metaverse, people who wanted to support their favorite artists, and gamers hoping to earn in-game assets. No longer was our industry purely about money; "cryptocurrency" interested only a subset of users.

It was only natural that the community began elevating the less-used but long-existing term *Web3* to describe a holistic technology movement with broad cultural as well as financial implications. The movement has its own particular values: one is that people should be autonomous and resilient, taking personal responsibility for the decisions they make on the web, safeguarding their own value rather than depending on corporations; another is that the systems for governing the web should be open and transparent, applying the same set of rules fairly to all participants. Some of these values are derived from the origins of the web. Others are still being shaped by newcomers to Web3. These are the next 100 million users, and they come from absolutely everywhere.

Web3 swept beyond finance and technology, and today, it's poised to intersect every industry. Similar to how every industry was affected by digitization, each will need to adapt to Web3. But the computer scientists and economics professors who got us *here*, to this level of crypto adoption and the starting line of Web3, won't get us *there*, to the next 100 million "mainstream" users. They will be joined by professionals leading Web3 teams inside their organizations, creative entrepreneurs and artists who understand how to engage with Web2 retail consumers, and investors who can read not only open source code on Github and academic whitepapers but also intricacies of human behavior. Far beyond a screen in the Geneva airport, this technology is truly on the verge of mass adoption. The people, projects, and companies that catalyze this next wave of adoption can expect vast rewards— and they won't all be engineers. There is a path for the brightest nontechnical marketing and business minds to lead with them.

My objective in writing this book is to empower fellow marketers to start building in Web3. I use the term *marketer* very loosely throughout to describe those who, like me, love telling stories and putting ourselves in other people's shoes, who care about how the things we create look and feel. We are businesspeople and strategists, artists and designers, community leaders and educators. As our movement touches wider audiences and more diverse industries, I believe minds like ours have a crucial role to play driving the next wave of Web3 adoption. To date, there exist few high-quality educational materials on Web3 for non-engineer readers. This is a shame because it deters some of the sharpest thinkers and most experienced professionals from entering the space.

In this book, I attempt to explain Web3 in the clearest way possible to anyone familiar with the internet. The most important point to understand about Web3 is that it's not a predetermined set of outcomes, but rather a substrate, a clay we can mold in our hands. Like an artist learning a new medium, we must understand our materials thoroughly before we can begin creating with them. For this reason, I spend the first third of the book tracing the history of how Web3 emerged from its origins in Web1 and Web2, showing that Web3 isn't a new set of ideas, but rather a novel realization of the original vision for the web. Then, I explain its key properties, the ones the readers of this book can use to mold their own Web3 systems. These include tokens, NFTs, decentralized applications (dapps), decentralized finance (DeFi), decentralized autonomous organizations (DAOs), and Web3-enabled metaverse worlds.

Nearly halfway through this book readers will encounter the first sections explicitly about marketing. These benefit from insights I've gleaned from reading numerous texts on marketing, and I cite many of my favorites throughout. No prior familiarity with marketing concepts is required to understand them, especially because many of my recommendations cut against the grain of traditional marketing wisdom. The marketing-focused chapters mostly detail the best practices and learnings my teammates and I have gathered over seven years as the first Web3 marketing team, first focused on Ethereum at ConsenSys, and now at Serotonin, the marketing agency and product studio I founded afterwards.

These chapters offer the reader—the artist molding the clay of Web3—examples from the work of the artists who came before, including ConsenSys companies, Serotonin clients, and other notable Web3 projects. The goal of these chapters is not to provide readers with blueprints or templates to copy exactly—though at certain points I do lay out specific strategies—but to shape an approach to Web3 marketing informed by past triumphs and failures. I've taken this approach so this book can remain useful for some time, even in a rapidly evolving space. These chapters include a design for a Web3 marketing funnel, suggestions of channels to use at each stage, and strategies for how best to use each channel.

Throughout these practical chapters, several refrains recur. At the risk of sounding like a broken record: the most important thing for marketers to remember in Web3, as it was in Web2 and long before, is to know their audience and to understand their product. A marketer's choice of product can make all the difference; if we hadn't chosen Ethereum, our team would surely have enjoyed far less success.

In the final practical chapters, I focus on strategies for building community, defining what that term means in Web3, and how it is the key to unlock sustainable, long-term growth for Web3 projects— giving them, I believe, the power to outcompete traditional and Web2 businesses over time.

Whether the projects you support end up beamed down from a Times Square digital billboard, spotted on the fashionable streets of the metaverse, or atop the rankings on Defi Llama, I hope every marketer has their own version of the experience I had in the Geneva airport. My hope is that by the end of this book, you will gain the confidence to seize the opportunity to shape the future of Web3. When you are ready use the code and instructions in the front of this book to claim your NFT membership to the Web3 Marketing community. There you will find more educational resources and training programs, as well as mentors and potential collaborators. I look forward to meeting you there, and I hope you enjoy the book.

PART

1

What Is Web3?

1

The Evolution of Web1 and Web2

IT'S NATURAL TO be skeptical of claims that a new technology will change everything. I was born in the early 1990s in a moment of techno-optimism. The recent fall of the Soviet Union had Americans like my family convinced that our system was basically correct, and that with this system in place, our society was ripe for innovation. The scale that new innovations could reach suddenly seemed infinite, thanks to Clinton-era globalization introducing low-cost production and efficient transnational supply chains. There was no doubt technology would progress linearly, if not exponentially, unlocking opportunities for prosperity and making them available to all. The invisible hand of social progress appeared to be steering us inevitably toward a more accepting world without borders. It was the end of history and the beginning of the greatest era in technology. For my cohort of friends, we expected a life of relative ease compared to our parents, our needs entrusted to kindly robots, our days spent whizzing over futuristic cities in flying cars, like those in "The Jetsons."

Suffice it to say I was disappointed. It was not the end of history, as Francis Fukuyama infamously proposed at the turn of the 1990s. Global hunger, poverty, and violence would continue shrinking, but my friends and I would contend with economic disasters, physical and psychological disease, and the collapse of trust in institutions and

3

between people, which is arguably the true wealth in a society. Not only that, there were no flying cars. The innovation that sprung up during my early life was in the field of bits, not atoms. In America, we would close our factories and retreat from building the kinds of things in the physical world that had captured my childhood imagination. We would instead turn inward: away from the kinetic world, toward the computer.

The breakthrough innovation of the 1990s would not be the flying car, but the internet. The global network of computers known as the internet dates back to the 1960s, and for decades its use was limited mainly to government and university researchers. At the turn of the 1990s, however, the internet became available to the public— and its growth exploded with the invention of the World Wide Web. The World Wide Web comprises a set of tools for creating and accessing documents (*web pages*) full of text, images, and other media, and most crucially, clickable links (also called *hypertext*) to other documents and resources on the global network. The development of graphical web browsers for personal and commercial computers made navigating the rapidly proliferating linked pages of the World Wide Web increasingly easy—and fun. Networking computers together in a web drew massive use to the internet by fostering conditions akin to Renaissance Florence: it became the place for information-sharing and access to like-minded people, but at a previously impossible scale. An ecosystem of content-rich, increasingly multimedia websites for governments, businesses, organizations, and individuals flourished, as first web message boards and later personal blogs provided tremendously popular platforms for regular people to publish and communicate online.

Although websites in this era were far more static and less interactive than what we have become accustomed to, the technology paradigm many have retrospectively come to call Web1 exponentially multiplied the amount and quality of information available on a global scale, democratized access to it, and radically lowered the barrier to entry to communicate and share knowledge. And the almost indefinite scalability of a piece of software on the web, which can in theory reach billions of people without additional up-front labor or production costs, introduced previously unheard-of opportunities for value capture that allowed the San Francisco Bay area to double down on its position as a global technology hub. Already home to the semiconductor industry, major players in personal computing hardware and software, and Stanford University, the area had the capital and technical acumen to begin supporting web companies. Web1 engendered an information revolution. Like other dreamers, I sighed about the lack of Jetsons-style flying cars—but encouraged by my savvy computer scientist father and neuroscientist mother, both of whom saw the future glowing on the other side of the computer screen, I begrudgingly adopted their optimism about the world of bits.

The original architects of the web were techno-optimists who believed open, networked systems would bring important social and scientific facts to light, helping break any grip on power by the nefarious or undeserving. Sunlight would be the best disinfectant. They intended the internet to be a place for discovering truth and having fun. The inventor of the World Wide Web, Tim Berners-Lee, and his collaborators took a philosophically decentralized approach to designing the first web architectures. The early web pioneer John Gilmore—a contemporary of Berners-Lee who started the Electronic Frontier Foundation, a nonprofit that promotes internet civil liberties—famously said of decentralized design, "The net interprets censorship as damage and routes around it."[1] Protocols like BGP (Border Gateway Protocol, allowing IP addresses to resolve to physical computers), TCP/IP (allowing information to flow between those computers), and DNS (allowing names to resolve to IP addresses) were all intended to be decentralized. Unfortunately, Berners-Lee and others didn't design for decentralization in the face of a malicious actor who wished to control the web, such as a government—and in fact, DNS, BGP, and other key web protocols are now largely under government control. There were technical limitations, too; early web architecture didn't allow for files to be stored in a decentralized manner, across multiple servers.

Despite these issues, from its underlying architecture to much of its early content, the internet met and exceeded the web pioneers' hopes. Many of them shared an ethos of utopian optimism with the first generations of hackers, whose exploration (sometimes malicious, often not) of computer networks and phone systems long predated the web. The Hacker Manifesto of 1986 envisions a borderless world of bits free from bias and discrimination, an even playing field where anyone with a computer and the necessary skills can fight to defend the underdog against the excesses of arbitrary power. Hackers took up arms against the functionaries of the decaying institutional order—the empty suits and clueless bureaucrats who couldn't even use a word processor—and sometimes they profited from it, breaking into insurance company databases, draining funds from faceless corporations, or simply enjoying free long-distance calling. These same computer geeks and their spiritual heirs would go on to run circles around their old-guard adversaries, not in the shadows as black- and gray-hat hackers, but on the open field of capitalist competition as developers of a new information economy, taking the highest-salaried jobs and launching phenomenally lucrative startups.

As it turns out, though, the real adversaries to the original spirit of Web1 were neither corporate suits nor government agents, but other web founders and their solutions for making the internet more widely usable—and wildly profitable. In contrast to the open, decentralized technology architecture of Web1, this crop of founders developed closed, walled-garden environments in the name of making the internet easier for everyday users to access. Internet service providers such as CompuServe, Prodigy, and, most significantly, America Online (eventually just AOL) essentially became competitors to the Web1 open standard, even as they formed the on-ramp to the internet for millions of regular users.

By the mid-1990s, many companies had begun providing consumers with internet access via a combination of software and dial-up connections. ISPs like AOL came to dominate, partly due to effective marketing (such as blanketing the country with free install CDs), partly due to user experience (UX) design maximizing ease of use. AOL and its ilk acted as mediating surfaces between the user and the actual open internet: a friendly voice welcoming you and letting you know you had mail, plus user-friendly design that made it easy to access chat and instant messaging, a fair amount of proprietary content that lived within an ISP's centralized system, and a web browser leading out to the open internet.

These holding tanks for users charged fees—at first by the minute, then in unlimited-use subscriptions—whereas the internet itself was intrinsically open and free. These companies were also able to collect substantial data on user activity, from user identity information, to shopping preferences as e-commerce began to emerge, to the actual content of user messaging. Without realizing it, consumers had begun to form habits with their internet use that involved certain trade-offs in cost, centralization, and privacy.

To a significant percentage of AOL customers, AOL was simply synonymous with the internet, and by the late 1990s, AOL was by far the largest ISP in the United States. Meanwhile innovation on the actual open internet continued chugging along. On the infrastructure end, advances in broadband technology meant that for regular consumers, not just institutions and businesses, the internet was increasingly something that was always on, not dialed into, and getting much faster. On the content end, just as broadband access from phone and cable companies were making AOL's phone numbers largely obsolete, the evolution of search removed the need for AOL's protective surface layer.

People were building more and more websites—but how was anyone supposed to find them? Early on, the web had *portals*, directory websites with lists of links. Yahoo! was originally called "Jerry and David's Guide to the World Wide Web" and was simply a list of links. Eventually, portals were replaced by search engines, including Altavista, Lycos, and Yahoo! itself. Google came later and eventually won because their PageRank algorithm was better than other search engines. Google Search made it so much easier to access the website a user was looking for, while avoiding dangers and distractions, that eventually there was no need for users to operate inside an enclosed environment to get the most out of the internet. Increased user volume coupled with excellent machine-learning algorithms conferred on Google the network effects that would make it a giant, with a sleek UX that was optimized into the single search bar.

By the early 2000s, social networking similarly provided users a mediated way to interact with the internet via a simplified UX. Although companies like WordPress would make it easier for users with no coding skills to publish their own websites, MySpace and other early social media sites enabled the noncoding majority to publish websites with the content and aesthetic qualities of their choice, at least within a standard template. From ancient astrology to the Myers-Briggs to BuzzFeed quizzes, humans have always craved learning more about themselves and finding ways to wrap that information in standard packages to project to others. Social networking offered us that treat at scale: one's profile, which in the kinetic world might reach thousands of people over the course of a lifetime, in the world of bits could reach millions or billions at once, if only they would pay attention. Facebook would introduce tighter standardization across its profiles and, with perhaps unintended brilliance, launch into one of the most aspirational market segments: Harvard students. Expanding at first only to other Ivy League universities in the United States, one at a time, Facebook achieved a veneer of exclusivity and attracted a long line of colleges and individual users waiting behind their velvet rope to be admitted to the club.

Google and Facebook are signature institutions of what some in the 2000s began calling *Web 2.0*, and what today we might simply call Web2. Web2, which is where most of us currently spend our time online, ushered massive numbers of users onto the internet. They came to shop, chat, and learn. With the advances in interactivity presented by higher connection speeds and evolving software and hardware infrastructure—including the explosion of mobile internet technology catalyzed by the 2007 arrival of the iPhone— these users were now creating, posting, and consuming increasingly massive amounts of content, day and night, wherever they were. The companies that onboarded them offered real value propositions and were effective at saving people time and money, helping friends stay connected, and disseminating knowledge. Instead of visiting multiple libraries and bookstores, we could instantly order any book we desired on Amazon. Information appeared at the click of a mouse, and Wikipedia garnered so many contributions (and revisions) it became as accurate as *Encyclopaedia Britannica*.[2] Decades after losing touch, without spending hours poring through dusty phonebooks, Facebook users reconnected with their childhood friends no matter where they lived. Even those who worried the most about sharing their credit card information became avid online shoppers as secure checkout systems emerged, thanks to a new generation of payment and e-commerce companies like PayPal and eventually Shopify. Each of these platforms was incentivized to onboard as many internet users as possible, so they used the data they gathered to make the UX they offered easy and accessible to everyone.

The founders of Google and Facebook—even more than the leaders of AOL and CompuServe before them—were young engineers, not the corporate and government drones who had been the laughingstocks of Web1 hackers. They were creating more scalable value through software and building on fundamentally decentralized technologies on the open internet. But as their organizations grew and took on investors from the now-booming Silicon Valley venture capital industry, they began facing commercial pressures. Namely, they needed to monetize their growing companies, and the best answer to the monetization question seemed to them and their investors to be found in advertising. Instead of companies charging users subscription fees, the users themselves would become the product—sifted, sorted, and sold to advertisers, who could use the data generated by users' online lives to target audiences with pinpoint accuracy. We would pay this price for the remarkable benefits of the internet today.

Google and Facebook were built on the open internet, rather than by creating an enclosed environment between the user and the internet as CompuServe or AOL had. But once users arrived on their websites, Google and Facebook could gather data on those users and hold it in centralized databases owned by those companies, as opposed to being visible and available on the open internet. These data silos were simply a new type of walled garden. A user wouldn't be aware of being tracked or having data about them collected. Instead, Facebook users would freely input their personal preferences (especially in the early days, listing favorite movies and books) and build a treasure trove of data with their activities on the platform, like the other profiles they chose to look at, whom they friended, whom they poked. Data collection was similarly unobtrusive to the Google user: Google's data collection engine hummed quietly in the background, feeding information about a user's searches, and eventually, once login, email, and the rest of the Google Suite were introduced, their personal identity.

What Google and Facebook learned about people personally— their interests, how they used the platform, and what they bought— powerfully informed the optimization of both platforms' UX to induce users into spending the maximum time on the platforms and maximally monetizing that time by exposing the users to advertising. For the first time, advertisers, or anyone at all who wanted to pay to reach people with any message, could pay to reach almost anyone, sorted by geography, age, income, preferences, or other identifiers.

By the 2010s, Facebook and Google had successfully disrupted the market for advertising, previously the domain of newspapers, magazines, and television stations, none of which could provide as direct access to users, as highly specified personal data about targets, or the low rates made possible by enormous scale. It was as if the newspaper boys of yesteryear had decided to unionize and demand an enormous share of subscription value because they were the ones with the direct access to the advertisement targets—what was termed in mediaspeak "the eyeballs," the scarce resource that Facebook and Google were actually monetizing.

The traditional media that depended on advertising were done for. The same for much of the digital media. Their death knell was becoming reliant on Facebook and Google for access to their own audiences, putting their distribution in the hands of new intermediaries they naively believed shared their interests. Today, advertising-based editorial media is almost completely dead, except for television, which is holding on longer. Other than TV, the only media companies that have persisted are legacy brands like the *New York Times* and *Wall Street Journal*, which can charge subscriptions, and those funded by billionaires or special interest groups for their own purposes that do not require the businesses to be profitable.

It's hard to say whether the young engineer founders of Google and Facebook were concerned they were causing harm. Both were executives of large companies bound by fiduciary duty to investors to grow the value of shares. Google was concerned enough to institute the unofficial motto "Don't be evil" (which it jettisoned in 2018) and has a reasonable argument to make that it adds positive value to billions of people's lives by indexing the world's information and organizing it to be discoverable. Similarly, Facebook's Mark Zuckerberg has insisted that connecting people with each other is inherently a social good, and indeed, users in disparate parts of the world who suffer from a rare disease can find moral support and raise funds in a Facebook group; a gay teen in a remote, politically conservative town can connect with others and learn they are not alone. Facebook and Google have undoubtedly led to lives being enriched and even saved. However, there are also strong arguments that companies such as Google and Facebook have caused significant ills, of the social, political, and economic varieties.

I will only briefly touch on the social and political havoc wrought by the technology giants that are now considered the problem children of Web2 to the degree these are relevant context for the emergence of Web3. From a social perspective, it's clear the interests of these platforms are fundamentally misaligned with those of the individual user. Let's imagine that a Facebook user wants to live a balanced life of sleeping, eating, working, using Facebook, and physically spending time with friends and family. But the Facebook corporation would rather they shirk their duties, constantly interrupt family dinner, and stay up later than is healthy at night—to spend more time on Facebook, so their company can then mine for data and monetize facing advertisers. There is also increasing concern about the mental health consequences of social networks. Intuitively, it doesn't seem particularly healthy for a girl in Poughkeepsie going through an awkward phase of puberty to be comparing herself to a supermodel in London on a platform that, far more effectively than a drugstore copy of *People* magazine, has been machine-optimized to fit her brain to the point where she can't look away.

Politically, too, fault can be found with both Facebook and Google. Massive audiences on those platforms have been reached by unidentified advertisers and possibly adversarial countries with the intention of manipulating public opinion. And paid ads are only one part of the problem. Bots and click farms that deal in engagements, content sharing, and participation in groups account for an unknown percentage of activity on Web2 social media platforms. The algorithms that govern discoverability and visibility have tended to privilege content that provokes the strongest emotions, as opposed to that with the greatest factual accuracy. Some thinkers like Yuval Harari worry that ever-optimizing machine-learning algorithms will grow to understand us better than we know ourselves, undermining the basic idea of democracy: that each citizen inherently has their own views, and the best way to govern is by drawing out and generally heeding what the majority thinks. If we spend so much of our lives being manipulated by algorithms to the point where our views are no longer truly our own, can the fundamental unit of democracy—the informed citizen—still exist?

While Web1 laid the foundation for an open, accessible information-sharing system, Web2 platforms are characterized by business models that are misaligned with users and sacrifice their interests in favor of shareholders'. From the perspective of early members of the Web3 community, the Facebook and Google business models epitomize the pitfalls of Web2. Many early Web3 founders conceived their projects as efforts to right the wrongs of a web whose original intentions were perverted into business models that aggregate personal data in walled gardens and monetize it in ways that cut users out of the value flow and benefit only the company shareholders. The rationale is one of economic fairness: if billions of individuals co-created the data product that a company is monetizing, should they not also benefit from the value generated by monetizing an asset—their data—that is fundamentally theirs? Web2 companies might argue that individuals agreed to terms of service allowing their data to be used in this manner. But just because something is nominally legal doesn't mean it's fair, and the economic ramifications of this unfairness are increasingly visible.

The Pareto distribution in economics refers to a system where winners accumulate more over time and losers lose more over time, leading to a widening divide between those with more and those with less, reaching an equilibrium where the top 20% hold 80% of the value. Many systems in nature, such as the allocation of water to plants through root systems, over time look like a Pareto distribution, with the healthiest plants getting healthier and the least healthy plants increasingly dying. This is sometimes called a "winner-take-all distribution" and sometimes the "Matthew effect," from the verse in the Gospel of Matthew, "For to all those who have, more will be given, and they will have an abundance, but from those who have nothing, even what they have will be taken away" (Matthew 25:29, New Revised Standard Version).

One of the most poignant effects of Web2 platforms is a widening Pareto distribution, whether in the context of how often a video is viewed on YouTube, a song is streamed on Spotify, or buyers are directed to a store on Amazon. Take the example of a musical artist on Spotify: if the main place to get access to listeners is Spotify, and hundreds of millions of listeners are on Spotify being served by algorithm and curated playlists, the small group of the most popular musicians will eventually receive an increasingly outsized share of total streams and therefore total economic returns from the platform. The bar for breaking into top playlists and attracting algorithmic attention is considerably higher than for attracting the attention of a local DJ. Then, if Spotify has a near-monopoly on demand for streaming, it gains the leverage to pay very little to creators for intermediating their access to audiences (on top of the relatively little that record companies have always left for artists after taking out *their* cut for intermediating access to audiences).

The economics for a top streaming artist may still be favorable even at those rates, but may be prohibitive to the emerging or niche artist, who is forced to look to other revenue sources (like touring, merchandise, subscription-style patronage) rather than profiting directly off the sale of music or streams. Streaming may generate more revenue for artists than the immediate prestreaming Web2 era, when users made digital copies of countless recordings available online for free—enriching both the networks who hosted the files (and advertised to downloaders) and the technology companies, like Apple, who benefited from users having access to free content to use on their devices. Nonetheless, Spotify's model does not currently allow many artists to enjoy a middle-class income, even as it benefits from the value contributed by the vast array of musicians who feel they must participate in the dominant music market of the day. Web2 has stretched the Pareto distribution for success creating all kinds of content, while funneling profits to those who preside over these hyperefficient digital networks.[3]

Whether fair or unfair, inevitable outcome or collective failure of imagination, an economic ramification of the Web2 business model has been to accelerate Pareto stratification: first, by helping small groups of Web2 investors and executives achieve extraordinary wealth by productizing their users; second, with their near-monopolies on flows of attention and demand that enable extraction of large rents to serve as intermediaries; third, by making it harder for creators on the middle and low ends of the popularity spectrum to get noticed and then remunerated for their work.

How did Web2 stray so far from the intentions of the designers of Web1? The answer lies partly in the market traction Web2 business models were able to achieve. Alphabet, the renamed Google, and Meta, the renamed Facebook, still have advertising to thank for 81% and 97% of their revenue, respectively.[4] But the reason these models were adopted in the first place points back to a core flaw of Web1. Although the foundational protocols of the internet such as DNS, BGP, and TCP/IP were designed to be fundamentally decentralized (albeit not in the face of malicious actors), the application layer that sat on top was inherently centralized in the servers of the businesses that created the applications. Without monetization possible at the more decentralized protocol layer, centralized business models needed to be created on top of decentralized architectures to capture value.

The original designers of the web actually had their own plan for monetization. Most people have clicked a broken link and stumbled across HTTP status code 404, "Not Found." However, no one has yet seen status code 402, "Payment Required," which is still described in web protocol standards as "reserved for future use."[5] The codification of this error in HTTP—the protocol at the very foundation of the web—demonstrates that the early web creators intended to build an internet-native payment layer that would sit on top of TCP/IP, at the HTTP layer (HTTP is the protocol that contains the response codes 404, 402, etc.), but they never got around to finishing it.

It's possible that history would have unfolded very differently had they realized their plan. That could have meant no PayPal or Stripe, none of the new intermediaries or banks extracting a value spread for online payments, or fewer advertising business models, because payment would already exist intrinsically to the internet. The idea of an internet with fluid, efficient native payments captured the imagination of the first Web3 builders. In building Web3, they weren't innovating something altogether new. They were restoring the web to its original vision, righting the perceived wrongs of history.

2

The Evolution of Bitcoin and Ethereum

On September 11, 2001, I was in the fourth grade at an Episcopal girls' school in Washington, DC. In the middle of that morning's English class, we were told a plane had crashed into the World Trade Center in New York, and that our parents would be coming to pick us up early; we were not yet aware of the attack on the Pentagon less than six miles away. When I got into the car with my mother, she was crying—something I had never seen before. "This might change everything," she said. "I'm not sure what happens next."

A confluence of historical circumstances in the first decades of the 21st century accompanied the shift from Web1 to Web2. These events would likely inspire Satoshi Nakamoto, the pseudonymous inventor of Bitcoin, to build the technological foundation of Web3. The first event was 9/11. After winning the Cold War, many Americans believed something like Fukuyama's thesis that history had reached its logical conclusion—namely, the Western order led by the United States, which, unlike previous failed empires and superpowers, would justifiably and basically peacefully dominate forever. After all, hadn't we introduced prosperity to the whole world by spreading democracy

15

and capitalism through globalization? Who could really be against that? Against *us*?

Of course, many dissenters had long said otherwise, but for many Americans, 9/11 shattered the myth of an untouchable last superpower and exposed wider cracks in the American mythology. These Americans woke up to a world of actors intent on toppling American economic, military, and cultural dominance, actors now obviously able to overcome their smaller numbers and limited resources with the asymmetric advantages of terrorism. Some Americans woke up to the possibility that many around the world had not benefited, but suffered under and on the road to American dominance—including many in America itself. Although the immediate aftermath of the attacks led to temporary feelings of sympathy and solidarity—American flag bumper stickers and "I heart NY" caps proliferated—what emerged was a climate of distrust.

The Patriot Act hustled through Congress suggested the ultimate distrust between the state and its citizens, who could be surveilled as potential threats by the US government even as that government purported to defend individual rights around the world. In turn, citizens struggled to trust that institutions had their best interests at heart, from full-blown inside job theories positing cozy relations between the Bushes and Bin Ladens, to a sense of the consumerist emptiness at the heart of President George W. Bush's pleas to Americans to preserve their way of life (and their economy) by going shopping again.

Joseph Lubin, one of the cofounders of Ethereum and the founder of ConsenSys, has said that 9/11—when he was chased down the street by a ball of fire on his way to his downtown Manhattan office—was the moment he realized something might be terribly wrong with the American system:

> Growing up, I and everyone I knew trusted society and the structures that we found ourselves embedded in. . . . 9/11 to me represented a loss of naïveté for many of us individually, and perhaps a loss of innocence as a society. The many learnings and disclosures in the decade or so following that day—from Lehman Brothers to the bailouts of too big to fail financial actors at the expense of taxpayers and homeowners—laid bare the complex and often unhealthy context in which we were actually embedded. Many

discovered that it was folly to trust those we implicitly felt had our best interest at heart. Politicians, bankers, pharma companies, insurers, car companies, health care organizations, they were all more focused on making a buck than on our well-being.[1]

History, in other words, was back in motion. It's instructive that Lubin immediately connects the shock of 9/11 to the financial crisis of 2008—which by that time was not the shock to him that it was to many. While the Bush Administration's credibility took further hits in the months and years after 9/11 thanks to the false intelligence used to sell the global "war on terror" (to say nothing of the administration's close ties to Enron or its incompetent response to Hurricane Katrina), the American economy recovered—partly thanks to novel financial products engineered by titanic financial institutions, and partly from a new wave of software companies such as Facebook, Google, Amazon, and Oracle, which grew into ambitious mandates during the early 2000s.

Gone were the noisy snake oil salesmen of the 1990s dot-com bust, when an overly exuberant Silicon Valley sprayed billions into Pets .com and their ilk, betting that the consumer internet would catch on faster than reality ultimately allowed. The new tech giants had learned that not every internet company would necessarily succeed, even if web use grew by over 2,000% yearly. They brought a maniacal focus to their search for product-market fit and monetization, the latter mostly through a mix of data collection, intermediation fees, and advertising. Facebook's motto was "Move fast and break things," and true to their word, that was exactly what the Web2 tech companies did. But it was the financial sector that would first appear to be broken.

The financial institutions' appetite for risk won them fortunes in the 1990s until the tech bubble burst at the end of the decade, then again from 2002 to 2008 when the market recovered. When the market reversed in 2008, the public was forced to bear the losses. A financial system most people hardly understood suddenly crystallized in the popular imagination as a machine for insiders to privatize gains and socialize losses. As Western governments bailed out the institutions deemed "too big to fail" and neglected to jail their executives, it was not surprising that major populist movements on both the right (the Tea Party) and the left (Occupy Wall Street) rose up in the United

States to protest a system that seemed rigged. For all their significant differences, these massive organizing efforts both focused largely on their own perceived lack of power in the current system and the unrealized power in numbers they believed would, once catalyzed, put them in control of that system, to defang or dismantle or redistribute it as need be. Neither movement had any idea, however, that an even more fundamental challenge to that system had just been established, quietly, in an unidentified corner of planet Earth.

In October 2008, a pseudonymous individual or group calling themselves Satoshi Nakamoto published to the internet the original Bitcoin whitepaper, titled "Bitcoin: A Peer-to-Peer Electronic Cash System." There had been previous attempts to create large- and small-scale digital currencies outside the jurisdiction of national governments. They had largely failed due to lack of adoption or were snuffed out by administrations guarding their right to unilateral control over the currency used in their countries.

The Bitcoin whitepaper described a currency enabled by a *distributed ledger*, laying the foundation for blockchain technology.[2] Here's how it works: many networked computers all over the world (called *nodes*) each maintain identical copies of a continuously updated ledger. You can think of it logically as storing *addresses* (that is, unique strings of numbers and letters), how many bitcoins each address controls, and the bitcoins each address has transacted in the past. For a full technical treatment of how this works, using what are known as UTXOs (unspent transaction outputs), see *Mastering Bitcoin* by Andreas Antonopoulos (2014). Each address is associated with a *wallet* secured by public-key cryptography: while bitcoins are sent to and from the publicly visible address, the owner of the address/wallet verifies access to the wallet (and authority to transact with the bitcoins inside) by using a *private key*, a separate string of numbers and letters that it behooves the user to keep secret, similar to a password. Though the term *wallet* implies that it stores the actual bitcoins, a cryptocurrency wallet actually refers to these public and private keys—or, by extension, the hardware or software involved with securely storing and using these keys—for accessing and transacting the coins associated with a particular address. The bitcoins themselves exist only as entries on the ledger.

Each time the Bitcoin blockchain processes a new *block*—a new package of transactions that took place over a given period of time that now need to be added to the ledger, plus cryptographic data that connects (or chains) the block to previous blocks (hence the term *blockchain*)—the nodes on the network update the state of their ledgers to reflect the new reality. Changing the data in a block is nearly impossible, as it would require changing all the blocks that come after it in the chain, and that would require consensus among all the nodes in the network. The network protects itself from malicious attempts to manipulate the ledger in favor of any participant through the mechanism of block *mining*. With each new block, computers running nodes of the network, known as miners, use their computational power, known as *hashpower*, to solve a difficult math problem. They ensure the security of the network by ensuring the block solution requires so many hash computations that it is extremely hard to reproduce and therefore unlikely to be falsified for any significant number of blocks. For this service, miners are rewarded by bitcoins from that block, known as *block rewards*.

Any computer that solves for an incorrect state of the ledger—for example, attempting to claim access to more bitcoins for a particular wallet than it actually holds—would be outvoted by the others and as a result, would be ineligible to receive bitcoin rewards. This is known as a *proof-of-work* system for mining, because each computer does work in terms of contributing hashpower. Proof-of-work elegantly solved the tragedy-of-the-commons problem endemic to public goods that pits the interests of the individual against the interests of the group. Miners are motivated to secure the Bitcoin network, which benefits everyone, by the individual incentive of block rewards, brilliantly aligning the incentives of all network participants toward the common goal of an accurate ledger.

That is, of course, unless a miner or group of miners controls more than half the hashpower on the network. This is called a "51% attack," where an individual or group can act maliciously to prevent the network from functioning properly by double spending coins or preventing real transactions from being confirmed. It is a vulnerability of Bitcoin only in theory because it would be far too expensive for any single motivated actor to control more than 50% of Bitcoin hashpower; such attacks have, however, taken down many smaller networks. Because such a high proportion of participants would need to collude to manipulate the system as to be virtually impossible, the Bitcoin network can function in a way described as *trustless* or *trust-minimized*: no single participant needs to know anything about or trust any other participant.

Few would be willing to undertake a valuable financial transaction without the element of trust. The reason banks and other financial intermediaries in the traditional system are able to extract value in exchange for facilitating a transaction is that they provision trust to both counterparties. Each counterparty does not trust the other, but both trust the bank. The Bitcoin blockchain confers the same or higher degree of trust on a transaction as a bank, more cheaply, scalably, and automatically. Fundamentally, the Bitcoin whitepaper offered a sound design for the first-ever secure digital currency to operate automatically, governed by software rather than in a discretionary manner by a group of people. This was the beginning of Web3. The implications are incalculable.

The first-ever block of the Bitcoin blockchain, known as the Genesis Block, came online on January 3, 2009. Nakamoto had put their academic theory into practice and launched the Bitcoin network. With Bitcoin, any wallet holder could transact in bitcoins with any other in a manner that was completely peer-to-peer, without any intermediary like a bank in the middle. Web2 payment apps today, such as PayPal and Venmo, give the impression of allowing users to transact peer-to-peer. However, this is not the case: their UX masks the reality that the payment service, similar to a bank, intermediates the transaction and charges a fee for their trouble. On the Bitcoin network, there are truly no intermediaries. Users transact directly with each other. Though there are Bitcoin transaction fees, currently these are extremely low, because the block reward is high enough to ensure that miners keep mining. (Eventually, the Bitcoin network will rely on a fee to miners once all bitcoins have been mined.) This makes Bitcoin a cheaper way to bank.

Because the blockchain is censorship-resistant, one drawback is that a transaction sent to an incorrect address or sent under duress cannot be reversed. In a truly decentralized network like Bitcoin, there is no governing body with the power to discretionarily change the state of the ledger. This drawback can also be seen as a core benefit. Banks can choose to freeze accounts or transactions for their own business purposes or under pressure from governments, so that no value custodied in a bank is truly one's own, because it is perpetually subject to censorship. However, Bitcoin wallets are self-sovereign, directly holding their own value with their own private key, offering users complete control over their own property. In a world increasingly subject to surveillance, where not only communications but also financial transactions are closely monitored on the web, self-sovereign wallets are a solution for pseudonymous financial privacy and control of assets.

Furthermore, authorities with control over their government-issued currencies can discretionarily choose to increase the supply of the currency, perhaps to fund a war or popular social program, but ultimately devaluing that currency and risking inflation. This is exactly the scenario Nakamoto sought to prevent, not by hacking the US Federal Reserve and destroying the traditional financial system, but by standing up a viable adjacent system with transparent rules governed by code and a fixed supply of 21 million bitcoins that people can opt into as an alternative.

Nakamoto enshrined this intention in the Genesis Block itself. Sandwiched between lines of code, they inscribed several words, a headline from *The Times* of London: "The Times 03/Jan/2009 Chancellor on brink of second bailout for banks."[3] Little more needed to be said. Bitcoin would become the foundation of an alternative financial system, in which no group of individuals could decide behind closed doors to privatize gains and socialize losses like they did in the wake of the 2008 financial crisis. It was a beacon of hope from the ashes of a failing system, and its features of censorship-resistance, peer-to-peer transactions, and cryptographic security would make it immensely valuable.

News of the Bitcoin whitepaper and the subsequent launch of the Bitcoin network did not reach me at Columbia University, where I was a first-year student. I was not alone. A cataclysmic shift in the workings of humanity's economic foundation did not perturb the English or Statistics departments where I spent my efforts. A small group of enthusiasts discovered Bitcoin on hacker message boards while I toiled over regression analysis and the meaning of Proust's madeleine. Academics, hackers, and computer scientists formed a motley crew of skeptical, then increasingly assured, Bitcoin supporters, who famously encouraged one another to "hodl" (after a comical message board misspelling of an investor's intent to "hold") as bitcoin prices fluctuated. It was good advice, though the only person I knew directly who was rumored to use bitcoin at the time, exclusively on the dark web, was the undergraduate pot dealer.

In 2011, during my junior year, Bitcoin broke out beyond the domain of clever internet weirdos and computer geeks, away from its bargain-basement prices, and into the public consciousness, including mine, with the explosive bust of Silk Road, a darknet market facilitating the sale of drugs and other contraband items. The site's founder, Ross Ulbrich (aka Dread Pirate Roberts), was arrested and convicted of multiple felonies related to narcotics trafficking, money laundering, and engaging in a "continuing criminal enterprise," a charge usually reserved for organized crime kingpins; he is now serving multiple life sentences in federal prison.

As it turned out, this early Bitcoin use case went far beyond helping college kids access then-illegal marijuana, with Ulbrich additionally under suspicion for trying to arrange contract killings. But even more interestingly, numerous law enforcement personnel working on the case proved unable to resist a personal fascination with Bitcoin: federal agents investigating Ulbrich tried to siphon from the Dread Pirate's treasure chest of ill-gotten bitcoins, landing *themselves* in prison. And the case's California prosecutor, Katie Haun, has gone on to an illustrious career as a crypto investor with Coinbase, Andreessen Horowitz, and her own Haun Ventures.

After the Silk Road case, Silicon Valley venture capital dollars—unfazed by the media narrative about Bitcoin that dwelled on drugs, black market weapons, and murder—started pouring into the creation of bitcoin wallets and into exchanges that provided on-ramps from fiat currencies like dollars into bitcoin and off-ramps from bitcoin to fiat currencies. Price action followed, and bitcoin reached a height of $1,238 before dropping back down within days to under $700. The first of the famously volatile crypto market cycles passed that year, affecting a community of still relatively few members. It was a bumpy ride that saw 744,408 bitcoins—worth nearly $17 billion in today's prices—stolen by hackers from one of the largest exchanges, Tokyo-based Mt. Gox. But the Bitcoin blockchain was still in its infancy. Only 5 of the 21 million total bitcoins had been minted. Exchanges like Coinbase, that would go on to achieve massive scale, wouldn't be founded until 2012.

While venture capital money began luring entrepreneurs into building primarily bitcoin wallets, on-ramps, and off-ramps—relatively simple applications that often introduced centralized, custodial layers on top of the decentralized blockchain, destroying the benefits of self-sovereignty in exchange for user-friendliness—a small band of computer scientists embarked on a quest for yet more novelty. A Russian-born Canadian teenager named Vitalik Buterin was among them. He was a computer prodigy fascinated with Bitcoin and a cofounder of *Bitcoin Magazine*.

Buterin and others sought to make Bitcoin, fundamentally designed as a secure digital money system, do more than move bitcoin tokens around a network. They saw the blockchain and the alternative financial system it facilitated as a substrate for building new kinds of financial products and software applications. Not only could a money system be governed automatically and transparently without banks or central authorities, but it could also give rise to a host of financial, cultural, and creative applications thanks to its decentralized architecture. To make an analogy to Web1, Nakamoto had invented email; a host of dedicated computer scientists all over the world, including Buterin, wanted to invent the whole internet.

These scientists began by building directly on the Bitcoin blockchain in order to expand its capabilities. One group (including Buterin) developed Colored Coins, a solution that enables users to manipulate digital assets on top of Bitcoin, but which failed to get significant traction. Efforts to do more with Bitcoin persist to this day, notably Elizabeth Stark's Lightning Network, which proposes to speed up Bitcoin, enabling it to handle more transaction volume, and the Stacks ecosystem, which encourages building applications on Bitcoin. Eventually, though, Buterin decided not to build on Bitcoin, but rather to create a new blockchain entirely, called Ethereum.

Proponents of Ethereum—now the second-largest blockchain network—argue that efforts to build sophisticated applications on the Bitcoin blockchain are like asking a dog to walk on its hind legs; the system simply wasn't built for this functionality. Buterin presented Ethereum for the first time at the popular Bitcoin Miami conference in 2014; like Nakamoto, he also published a white paper. The new network he proposed would not only be decentralized, censorship-resistant, and trustless, with peer-to-peer transactions, cryptographic security, and self-sovereign wallets unlocked by private keys, it would also allow coders to upload and execute programs on the network for a wide variety of applications.

The Ethereum network is powered by the Ethereum Virtual Machine, which executes these programs—called *smart contracts*—stored on the Ethereum blockchain (for more on smart contracts, see the section in Chapter 3 on decentralized applications). Ethereum was not merely a system for individuals to send each other tokens, though it has a native token—*ether*—that works as an economic incentive for nodes to perform the requested computation on the network. Paying nodes to perform computation serves as an economic disincentive against overusing the network; these *gas fees* for each transaction are designed to try to prevent the network from becoming overclogged with transactions and slow, like Bitcoin had become. Ethereum was not an app; it was the proverbial app store, the substrate for building anything on the internet with a use case for internet-native money. If Ethereum realized its promise, it would restore the original vision of Web1 and far more.

3

The Evolution of Ethereum into Web3

ETHEREUM CHANGES EVERYTHING in a way that flying cars would not, though I'm still disappointed that the most important technological progress in my lifetime—Elon Musk ventures aside—seems to be in the world of bits and not atoms. I'm less excited today by flying cars, spaceships, and even advances in genomic medicine than by the possibilities Ethereum unlocks, because the former are fundamentally products or commodities. These can be novel and important, but as individual products being exchanged, they exist at the surface layer of the deeper economic system. Ethereum offers a new and different underlying foundation for building an alternative economic system with all the benefits of the traditional one, in addition to new benefits conferred by decentralization and the ability to build any economically and socially important application on top of the blockchain.

Bitcoin is fundamentally a system for moving bitcoins securely between wallets. Ethereum similarly allows users to move ether tokens between wallets. The difference is that this capability to act as a money system only scratches the surface of Ethereum's functionality. What is important to understand about Ethereum is that it's not a closed set of predetermined features; it is rather a substrate, or a moldable clay, for building anything that involves value in a digital context.

Since the Ethereum network launched in 2015, computer scientists, developers, governments, enterprises, philosophers, artists, and everyone in between have innovated on top of its platform in several key directions. But the capabilities of Ethereum are certainly not limited to the functions that have progressed the farthest today, and even these functions should be considered as tools in the toolkit or paints at the easel of anyone looking to build in Web3, as opposed to a finite set of options. I wouldn't be surprised if the most innovative use cases for Web3 have yet to emerge at the time of this writing in 2022.

To mix metaphors once again and return to that of substrate or clay to be molded, my goal is to offer nontechnical readers the information about the properties of this clay—its weight, density, color, and mineral composition—necessary to begin molding their own sculptures, and to inspire confidence by conveying an understanding of the canon of the best sculptures that exist today and how their creators made them. As a reader of this book, you can become one of the builders of Web3—the latest, decentralized, blockchain-based generation of the web—with the potential to evolve today's broken systems into something more transparent, efficient, fair, and inclusive for all. Whereas Web1 democratized information access and sharing and Web2 brought in massive use, Web3 holds the power to outcompete the outdated, extractive Web2 business models and usher in the healthy, open web imagined by its visionaries. It's in the hands of our generation to mold.

This journey of co-creation begins with understanding the key set of activities the Ethereum blockchain enables: tokenization, decentralized applications (dapps), decentralized finance (DeFi), decentralized autonomous organizations (DAOs), non-fungible tokens (NFTs), and Web3-enabled metaverse worlds. Along the way, where relevant, we will introduce key technologies: smart contracts, alternative Layer-1 blockchains to Ethereum and Bitcoin, and Layer-2 scaling solutions on top of Ethereum. These will be applied to real-world use cases spanning industries that include finance, infrastructure, energy, software development, arts, entertainment, and gaming.

Tokenization

A *token* is simply a representation of something else, often value. One can think of a US dollar as a token, because it represents a unit of value in a US currency presided over by the US Federal Reserve and Treasury Department. An IOU is a token that informally represents one person's debt to another. A token of affection can have financial value or none at all; it is a representation of someone's feelings.

Bitcoin is a token that represents a unit of value on the Bitcoin network. What sets it apart from the other types of tokens mentioned is that it sits on top of a decentralized blockchain. Similar to fiat currency, bitcoin and ether tokens are fungible—that is, their nature is such that any quantity is interchangeable with and not meaningfully different in any way from any other equal quantity—durable, and, increasingly, accepted. Tokens that live on a decentralized blockchain confer additional benefits beyond fiat money: they can be sent frictionlessly peer-to-peer without banks or other intermediaries. They are easier to store in large quantities than physical cash and are able to be stored under the control of the user, as opposed to digital currency in a bank that could choose, be forced, or in error freeze the account or drain its funds. Also, thanks to the blockchain, they are more open and transparent than even digital fiat currencies, and they can be fractionalized into extremely small quantities, which is especially helpful for some applications like micropayments.

Unlike physical dollar bills—which can be transferred untraceably at the coffee counter or in a briefcase—every time a bitcoin or ether is sent from one wallet to another, there is a permanent, immutable record of the transaction stored on the blockchain, which updates the state of the ledger (to put it simply, who now owns what) with each new block. Though digital dollars can be traced, and massive servers maintain the state of the ledger for actors such as banks, multinational corporations, and sovereign states, they are subject to hacking, manipulation, and errors. In a transaction between two actors, one actor can theoretically claim that the state of the ledger is x, and the other can claim it is y. Because each is maintaining its own ledger on its own servers, it can be difficult to discover the true state without a disinterested third party providing a ledger that both parties can agree to trust. Banks often serve this purpose in transactions in exchange for a percentage of the transaction value. But banks also have their own interests, and certainly their own fees, so a ledger built on a decentralized blockchain, maintained automatically by a network of nodes all over the world rather than discretionarily by a group of people, can serve as a more trustworthy third party, often at a very low cost, obviating the need for a bank. The blockchain ledger serves as a source of truth, accurately tracing the movement of tokens.

Fractionalization also sets blockchain-based tokens apart from fiat money. Unlike a physical dollar, a bitcoin can be split into extremely small units. While a dollar can be split into one hundred cents, a bitcoin can be split into 100 million pieces, called *satoshis*. One ether can be split almost indefinitely: the smallest unit of ether is called *wei*, and one wei is equal to one quintillionth (0.000000000000000001) of an ether.

Fractionalization means that no matter how valuable tokens become in dollar value, it will always be possible to transact using them at any price point valued in dollars. If a single bitcoin is worth $50,000, but a user wants to send $50 of bitcoin to another wallet, they can easily send fractionalized bitcoin. Though digital dollars can be split into small units as well, most transactions involving these denominations of currency, such as in business models based on micropayments, don't make sense considering the fees for intermediating fiat transactions. Tokens that enable peer-to-peer transactions with zero or extremely low intermediation fees unlock the logic of micropayment models in industries like media and remittances, and enable a global marketplace of fluid and frictionless transactions.

Similar to fiat currency, but unlike previous attempts at digital currencies, blockchain-based tokens like bitcoin and ether solve the internet's notorious "double-spend problem." On the internet, as on any personal computer, it is easy to create nearly infinite copies of just about any file or digital object. This leads software products to be efficiently scalable but presents a problem for value systems—and money systems in particular—where price is subject to supply-and-demand dynamics. If an infinite amount of currency is circulating—that is, if there is infinite supply available on the market—no matter how high the demand is, the price will be reduced to zero. Being able to infinitely copy and paste audio files upended the music industry; being able to infinitely copy and paste currency would be a bigger problem.

The blockchain elegantly solves this problem, because tokens aren't files on anyone's computer: they are units of value in accounts whose balances are a matter of synchronized consensus across a decentralized network of computers. New native tokens to that blockchain are only introduced with each new block and may not be generated otherwise. New tokens are minted automatically according to a mechanism that is public information. As the ledger updates with each new block, the new state of the network proves that no additional tokens have been minted. When a token like bitcoin or ether is spent—transferred from one holder to another—it can only be spent once, according to the rules of the network, with a record of the transfer distributed across the network that is all but impossible to dispute or alter. Trying to spend that token again would yield the equivalent of an "insufficient funds" message, not from a centralized bank, but from the decentralized blockchain network that keeps track of account balances.

This means that a lot is riding on a network's code, and errors in code could prove to be a network's downfall: some newer blockchain-based tokens—like those associated with Solana stablecoin protocol Cashio and DeFi platform Acala—have suffered from "infinite mint" attacks to their code bases, in which hackers discovered vulnerabilities that allowed them to mint additional tokens outside of the intended distribution scheme. Larger networks like Bitcoin and Ethereum that have existed for many years with billions of dollars' worth of tokens on chain have been so battle-tested by would-be hackers, drawn by the massive prize an exploitable vulnerability would represent, that most believe the chances of such a hack are infinitesimally small. Ultimately, Bitcoin and Ethereum are some of the first successful networks to produce digital currency with the core benefits of fiat currency (durability, fungibility, acceptability) and the added benefits of the blockchain (traceability, a ledger governed by code instead of discretionarily by people, fractionalization down to small denominations).

Ethereum, however, is more than a digital money system. It is a computing platform that anyone with the appropriate skills can use to develop their own tokens and ascribe to them any functionality. After Vitalik Buterin published his white paper and introduced Ethereum at the 2014 Bitcoin Miami conference, he and a group of collaborators who gathered around the project (some of whom would be named as cofounders) formed the Ethereum Foundation, a nonprofit foundation based in Switzerland whose job would be to develop and launch Ethereum. There was precedent for open source software platforms being managed by foundations, notably the Linux Foundation. Its goal was not directly to make a profit, but to build, launch, and continue to develop the software of the Ethereum network. In 2014, the foundation raised $18.3 million in a presale of ether tokens—priced on a sliding scale, with a final value of 1,337 ETH per 1 BTC—and used the funding to build Ethereum.[1]

By 2015, one of the Ethereum cofounders, Joseph Lubin, had amicably split from the Ethereum Foundation to found a for-profit software company called Consensus Systems (ConsenSys) with the goal of bootstrapping the adoption of Ethereum by building essential tools, services, and companies. Both the Foundation and ConsenSys continued innovating on Ethereum technology, at first working closely in alignment, with many shared employees and projects.

ConsenSys engineers contributed to building the ERC-20 token standard, which codified an approach to creating new tokens on top of the Ethereum blockchain that could be programmed with their own functionality. The standard made it easy to tokenize just about anything. As a result, most of the tokens on the list of the top crypto tokens by market capitalization today are ERC-20 tokens. Here are a few examples of how tokens are being used:

- **Security tokens** are programmed to function like any other security such as a stock or bond. In the United States, a security token must be sold only to accredited investors or else be registered with the SEC as a security, like a public stock, to avoid running afoul of US securities laws.
- Tokens can be used as **access keys** to use a product or service, like in an arcade where gamers purchase tokens and use them to play a particular game.
- Ownership of **governance tokens** represents the ability to vote in a group, decentralized autonomous organization (DAO), or company decision directly on the blockchain, with whatever degree of governance power is programmed into them.

- **Membership tokens** are used to grant membership to a club, such as an online social group, or a rewards club where holders can benefit from discounted access to products or services, the digital equivalent of a Costco or Sam's Club membership or airline miles.
- **Rewards tokens** are used to incentivize desirable behavior, like contributing to an open source code base, creating content, participating in a bug bounty program to help battle-test an early dapp, or referring a product or service to others. They can also be used to retroactively reward past behavior, such as being an early adopter of a certain product.
- **Liquidity provider (LP) tokens** are a ubiquitous type of token that represent injecting liquidity (that is, tokenized value) into a pool of tokens, often in a decentralized exchange (DEX), allowing the DEX to serve its users by giving them access to supply of a particular token. Users can access this supply through an automated market maker, which essentially automates the process of matching supply to demand for a given token so users can trade tokens directly without intermediaries. Users receive the LP token in return for providing liquidity; it represents their claim on the liquidity provided. LP tokens can also lose value as a result of their underlying liquidity being provided with a disadvantageous set of parameters. LP tokens have proven to be part of a lucrative mechanism in many instances, and their use helps DeFi function smoothly.

Although there are infinite use cases for tokens, what they offer is always the same at the core: putting structure (like the outer membrane of a cell) around something of value, thus enabling this value to be represented in an economic system and exchanged for other tokens that have economic value. The result is an economy with a larger total amount of value, thanks to existing value previously unrepresented in economic units now capable of being priced and exchanged. At the same time, not all value is financial, and tokens do not necessarily have any value in dollars. For example, a parent could sell tokens to raise funds for a child's birthday party. They could determine that a token costs $5, and each token allows one guest to attend. At the door they could verify on the blockchain that each guest held at least one token in their wallet. After the party, supply-and-demand dynamics would determine whether the party-access tokens still in the parents' wallets had any financial value. Chances are, while that token was worth $5 to the attending families, not many people will be interested in buying a random child's birthday party token for $5, or even $0.05—unless, of course, it turns out to be the fifth birthday party for a child who becomes famous later in life, in which case someone might at some point be very interested in buying such a token as a piece of digital memorabilia.

Many companies and individuals who produce tokens do so with the intention of capturing financial value with them—that is, that they can be priced in terms of other currencies and tokens and exchanged. However, the fact that they used a blockchain-based token doesn't automatically make the token valuable; only supply-and-demand dynamics can make a token valuable. This is why, similar to any other product, it's essential to consider the value proposition of a token, thoughtfully design how it captures that value, attempt to stimulate demand from the market, and course correct until product-market fit is achieved.

Anyone can originate a token, even with limited coding capabilities. Token-building applications that use the ERC-20 token standard and others, built on different blockchains, are easy to find on the internet. Once minted, tokens can be distributed from their point of origination through a number of mechanisms. They can be sold by their originator at a fixed starting price or by an auction that determines the market price. Often this happens on a centralized token launch platform such as CoinList, or on a centralized exchange like Binance.

The original term for this type of launch was *ICO*, or *initial coin offering*. Tokens can also launch directly on the aforementioned peer-to-peer marketplace known as a decentralized exchange, or DEX. This is sensibly enough called an *IDO*, or *initial DEX offering*. Tokens can also be sent from where they are first generated to any group of designated wallets. This is called an *airdrop*. A new Web3 product may airdrop its token to all the wallets of users who have historically interacted with another product, because all past transactions are visible on the public blockchain. Any action on-chain might represent a market segment relevant to the product.

They may wish to do this to target a particular cohort that they believe would be interested in using *their* product, as top-of-funnel marketing. These speculative airdrops, however, can distribute tokens to wallets of users who don't care about the tokens and will immediately sell them. For this reason, some projects choose *retroactive distribution*, or distributing tokens based on past behavior to users whose wallets have a history of interacting with their product, proportionally to how much they used the product, or by whatever metric they choose. Chances are, if a person has used the product before, they care about it and may prefer to hold their tokens or buy more to access whatever utility (such as governance, discounts, future rewards, platform use, sharing in upside) has been programmed into the token. Wallets can receive their retroactive distributions by airdrop or by the user proactively connecting their wallet to a website to demonstrate their eligibility and then claiming their tokens (a *claimable*).

Just as there are indefinite possible designs for tokens, there are also myriad ways to distribute tokens to wallets, with various advantages. Later on, we will weigh these alternatives in detail. The focus of these sections is on what is technically possible to achieve using the tools of Web3. That being said, not all token designs and distribution plans described are necessarily legal in every jurisdiction worldwide. We are starting to understand what this technology is capable of, but the law is behind the tech, and regulatory frameworks for Web3 are still nascent and evolving in most major markets. Traditional and Web2 companies, Web3-native projects, and even decentralized protocols with pseudonymous teams (more on this later) alike are well advised to seek counsel from lawyers with experience in blockchain and crypto before launching a token and to weigh the benefits and risks for themselves.

Once a company or project launches a token, it can be held accountable in courts of law where it is domiciled or potentially anywhere in the world the software is used. Less formally but often just as powerfully, the team behind a Web3 project is also held accountable by its community of token holders and by the crypto community at large. Even if there are no legal ramifications, teams that deceive their communities can have their reputations permanently ruined in the Web3 industry, such that they are unable to continue working in the space. Unfortunately, not every project takes this responsibility seriously. This, combined with often significant financial incentives to launch tokens, has led to a proliferation of scams in early Web3 that has sadly damaged the reputation not only of individual perpetrators but also the industry as a whole.

Newcomers considering working in Web3, however, are missing out on an opportunity if they throw out the baby with the bathwater. The early web, too, was crawling with scam artists trying to make a quick buck during moments of investor exuberance about the technology. That didn't mean the technology itself was without merit; to the contrary, it would radically transform society and business. It just meant that investors and consumers needed—and still need—to inform themselves and do their own research to separate the wheat from the chaff, and that serious builders and entrepreneurs needed to go the extra mile to prove their credibility to consumers and potential backers. Valuable companies on the web would emerge over time, while the market and legal frameworks that were eventually established would weed out bad ideas, scammers, and lawbreakers. This same evolutionary process is happening today in Web3.

Those who care about the future of Web3 should understand that their actions today, in a still immature Web3 ecosystem, will reflect on others who are trying to build using these tools. Bad actors make it more likely that the regulation that emerges will be broad and harsh, stifling innovation in a new sector of the economy—one of the few, crucially, that is genuinely growing, especially in the post-industrial West. They also slow the progress of Web3 by making potential users hesitate to adopt even high-quality Web3 innovations. However, because we are at this early stage, the builders, investors, and creators working in good faith to solve important problems using Web3 can also make a disproportionately *positive* impact on the credibility of Web3 at large, hasten its adoption, and invite thoughtful approaches by regulators who are better able to see the benefits that Web3 has to offer, and therefore more interested in protecting them. The future of Web3, in this regard, is in the hands of the readers of this book.

Importantly, not every Web3 project needs to have an associated fungible token or NFT. Many of the most widely used products do not, such as MetaMask. Most of the Web3 marketing strategies in subsequent chapters of this book apply to projects with and without tokens, which can be considered an optional—and in certain cases, advantageous—tool in the marketer's toolkit for building incentive alignment engines using the substrate of Web3, and forming their Web3 communities.

Decentralized Applications (Dapps)

With Ethereum and other newer blockchain networks, transactions need not be performed manually. Individuals can access their wallets with their private keys and send funds wherever they would like; but also, thanks to smart contract technology, applications have logic programmed in that disburses tokens when specified conditions are met. *Smart contracts* are neither smart, nor are they contracts. They are basically if-then statements expressed in code. For example, a smart contract could be programmed to disburse one ether from a particular wallet to another wallet if the weather in San Francisco according to Weather.com is over 60 degrees Fahrenheit at 9 a.m. Pacific Standard Time this coming Tuesday. In this case, Weather.com (paired with a protocol like Chainlink that would actually pull the off-chain data into use on the blockchain) is part of the *oracle*, the name for a provider of real-world data in a smart contract that determines how it executes.

No one needs to manually check Weather.com or use private keys to access the wallet holding the tokens. The smart contract executes automatically if the conditions are met. Extrapolate this outward to many different pieces of business logic strung together by various contingencies (if x happens, do y) and you have a software application that can be programmed to do just about anything involving tokens. Dapps are webs of interconnected smart contracts designed to serve a particular purpose. They can be used for borrowing and lending, for betting on anything from a sports game to a presidential election to the weather in San Francisco, or for accessing any sort of tokenized product or service. Tokens are used inside decentralized applications as payments, proof of ownership or authenticity, or for a host of other use cases. What makes a dapp different from a Web2 application is simply that it is built on Web3 decentralized architecture.

Decentralized Autonomous Organizations (DAOs)

By July 2015, a year after the crowdfunding presale, the Ethereum network was live—but without key developer tools and infrastructure, no one would be able to use it. The Ethereum Foundation raised the funds in bitcoin that it required to launch, but there needed to be a way for the group of interested parties—the foundation, the scientists and engineers contributing to the Ethereum project, and the investors who participated in the original ether presale—to fund their creation directly on the blockchain from a designated pool of ether, without needing to reveal their identities to each other or institute a traditional governance process like that of an investment committee at a traditional fund. The blockchain-native mechanism they built was called the DAO (short for *decentralized autonomous organization*), which would deploy a pool of ether that was sent by people who invested in the DAO through its smart contract. The DAO was a dapp built on Ethereum that allowed members to vote on how the ether in the pool was deployed, in proportion to their participation level. It was built on the decentralized blockchain and functioned autonomously, deploying capital according to on-chain votes by its members.

Unfortunately, the DAO turned out to have a critical security vulnerability, and in 2016, it was hacked. Exploiting a flaw, the hacker stole 31% of the ether in the DAO, which was at the time about 5% of all ether in circulation.[2] The hack made headlines. It was one of the first times Ethereum appeared in the *Wall Street Journal*, and was its very first cover story.

The DAO hack happened just as I was getting to know the early ConsenSys team. I was at a backyard barbecue in Brooklyn with Sam Cassatt (whom I married three years later) when his grandmother called, worried sick. She had seen the cover of the *Wall Street Journal*. Some were afraid that the DAO hack would be the death knell of Ethereum. As a marketer, I remember thinking, *At least the word* Ethereum *is making it into headlines*. Even though the coverage appeared negative on the surface—after all, something bad had happened to Ethereum—it also alerted general readers to the extraordinary amount of value on the Ethereum blockchain, so much that a hacker would go to all the trouble to steal it. The coverage not only introduced Ethereum to millions of people, it also subtly seeded the idea that although possibly flawed, Ethereum was worth their attention. In the wake of the DAO hack, and after a number of tense conference calls, most of the Ethereum miners decided to update the code base running on their nodes to an updated version of Ethereum designed by Buterin and his colleagues at the Ethereum Foundation that included a fix for the vulnerability. A minority group forked off and decided not to update their code, but instead to continue running the original Ethereum code base, calling their blockchain fork *Ethereum Classic*.

DAOs would become understandably unpopular in the following few years, although the error was in the particular execution of the Ethereum DAO, not in the general concept of a DAO. It wouldn't be until 2019, in the depths of a long market pullback called *crypto winter* in which the prices of ether, bitcoin, and a number of smaller ERC-20 and other tokens termed *altcoins* crashed from all-time highs to devastating lows, that Ameen Soleimani, formerly one of the earliest ConsenSys developers, reintroduced the DAO trend with Moloch DAO.

Moloch was special because, instead of simply being a single DAO, it introduced a repeatable framework for building more DAOs—just as the ERC-20 token standard had lowered the barrier to entry for creating a new token, leading to a wild proliferation of new tokens. Many DAOs today use the Moloch smart contracts, as well as other templated DAO smart contracts from subsequent DAO tooling projects, like DAOStack. This is the power of open source development.

The ConsenSys portfolio company OpenLaw (now called Tribute Labs) would pioneer the LAO (short for *limited liability autonomous organization*), a template for building a DAO in compliance with US or international laws. Founders like Soleimani and OpenLaw's Aaron Wright figured the reputation of DAOs had been unfairly tarnished by the DAO hack, and the bottom of the bear market was the perfect time for crypto market participants to pool their resources together to co-invest in new innovations.

DAOs have a number of use cases today, investing being the most common. By forming a DAO on-chain rather than a hedge fund or venture fund, DAO founders (or *summoners*, according to the colorful vocabulary proposed by Moloch) could avoid the fees and headaches associated with registering a fund, structure a simple capitalization table, and instate logical governance (typically, weighted voting by DAO members according to their level of participation). Moloch-style DAOs offer an easy formula for admitting new members: popular vote of the existing DAO. Other types of DAOs, such as the one that governs the Maker DeFi protocol, simply require holding its governance token for admission.

In a DAO, unlike a blockchain, decisions are not made automatically, but discretionarily by a group of humans. DAOs do not remove the human element from decision-making. Instead, they organize it, enabling groups of people to quickly and efficiently spin up a structure that enables them to deploy funds on-chain, programmed with whatever governance rules they choose. Sometimes individuals can even participate pseudonymously, without revealing identity information. The Moloch DAO contracts, and many DAO tools that followed, also enable members to withdraw from an investing DAO whenever they choose, with whatever portion of their contributed funds remain unallocated. Moloch named this action the *ragequit*; others simply call it *withdraw*. In other DAOs, such as MakerDAO, members need only sell their associated governance tokens, in this case MKR, in order to leave.

Whether or not a DAO is a LAO with an associated LLC or other entity structure, it is usually far easier for a group of people to get started allocating capital through a DAO than a traditional fund, especially if they are deploying and investing in Web3 native assets like tokens. As a result, DAOs today that focus on activities such as NFT investing hold billions of dollars in assets. Flamingo DAO, one of the LAO DAOs formed with OpenLaw, was one of the first NFT DAOs, and its success helped spur other popular NFT DAOs like PleasrDAO and BeetsDAO. It's important to note that despite their popularity, not all DAOs or even LAOs are necessarily legal today in every jurisdiction and may be subject to existing laws concerning investment vehicles, as well as emerging regulations on crypto and Web3.

But not every DAO needs to focus on for-profit investing. Kimbal Musk's Big Green DAO and the Endaoment use the DAO structure to aggregate funds for impact and charitable causes. DAOs can form with the goal of purchasing a single asset, such as ConstitutionDAO, which formed in 2021 to attempt to purchase a rare copy of the US Constitution for sale at Sotheby's. Its members were each issued a fungible token called *PEOPLE* proportionate to their level of participation. Even after the DAO lost the auction to Citadel hedge fund titan Ken Griffin for $42.2 million, the PEOPLE token continued to have market value, as not all DAO members chose to withdraw, some because of the friction of converting their tokens into other tokens, but others because they continued to believe there was value in the group that had assembled itself on-chain. The idea that communities such as DAOs have intrinsic value apart from the value of assets they hold becomes fundamental in our chapters on Web3 community.

A fourth purpose for a DAO is to govern a decentralized protocol or dapp. Take, for example, DeFi protocols such as MakerDAO, Compound, and Aave. These protocols are gradually moving away from the traditional model of governance through a single centralized legal entity, and toward governance by their respective communities through a DAO structure. There are many benefits of decentralizing governance. First, the projects need to employ fewer people if a community of relative experts and power users are willing to join and participate. Second, many Web3 projects share a philosophy that the stakeholders in a given project should be the ones, proportional to their stakes, making the decisions that lead to its success or failure. Delegating governance powers away from the core team and to the community of stakeholders does not mean robbing the project of its economic upside, because the project may still hold many of its own tokens. As long as the project has proper token economics (*tokenomics*), those tokens will increase in value if the project is successful. By aligning economic upside with its stakeholders in a single community of token holders with rights to participate in governance, the governing DAO of a DeFi protocol can aggregate higher-quality decision-making than it would have on its own.

There are also legal reasons—such as avoiding running afoul of the spirit of regulation—why certain protocols would wish to start out completely decentralized and pseudonymous, without a legal entity, or to gradually decentralize over time to the point where a legal entity can be removed. If a protocol is simply operating autonomously on the internet, governed by thousands or millions of wallets, in theory at least, no single centralized entity or individual can be held responsible for its activities. Legal tests of this theory are already emerging, their outcomes likely to be determined by the degree of decentralization and pseudonymity such projects and their contributors can maintain. But dapps and decentralized protocols will choose the route of DAO formation and progressive decentralization regardless in order to harness its other benefits.

Non-Fungible Tokens (NFTs)

Tokens and DAOs have always been possible to build on Ethereum, from the day it launched in 2015. However, they became popular once developers created technical guidance for constructing them, notably the ERC-20 token standard and the Moloch DAO smart contracts. Similarly, non-fungible tokens, or NFTs, could always be built on Ethereum, but the introduction of a new token standard for NFTs called ERC-721 in 2018 ultimately catalyzed their adoption. The story of why the ERC-721 standard was built would amuse even the creators of Web1 with their fun and exploratory vision of what the internet could be, though it briefly befuddled early crypto.

In 2017 a Canadian gaming studio called Axiom Zen launched CryptoKitties, a blockchain game built on Ethereum. CryptoKitties was a series of NFTs. Each NFT was a cartoon cat with different "cattributes," including but not limited to googly eyes, fluffy fur, calico pattern, tiny horns, capes, and wings. Players could "breed" two cats with each other to create a new generation of CryptoKitties NFTs. The NFTs were considered valuable based on their cattributes, some common and some extremely rare. The small but growing Ethereum community was enthralled. Many of my colleagues at ConsenSys kept a tab open on their computers at the office for trading and breeding CryptoKitties, and would hop back in to play between meetings.

So great was the enthusiasm for these cats that demand congested the Ethereum network with too many transactions, making the network slower in handling other, possibly more business-critical types of transactions and driving up the prices of gas fees. Developers from leading ConsenSys-backed projects like MetaMask, the now-ubiquitous Web3 native wallet, and Infura, the Ethereum infrastructure provider, were called in, along with developers from the Ethereum Foundation. They collaborated on a series of optimizations and scaling solutions to help unclog the Ethereum network and ensure it wouldn't happen again. Meanwhile, rare CryptoKitties NFTs were selling for six figures. In 2018 at the second Ethereal Summit, the Web3 culture-focused conference series I cofounded during my time as ConsenSys CMO, a CryptoKitty that sold in a charity auction for $140,000 was written up in the *New York Times*.

Cats on the internet have always been captivating, but Crypto-Kitties were far more valuable and engaging than any other cat image or video online. The reason is that they were NFTs. Although the tokens discussed previously, such as bitcoin and ether, are fundamentally digital money, with the same properties as money (durable, fungible, acceptable) in digital form, NFTs are intentionally not fungible. Any single bitcoin is the equivalent of any other bitcoin, just like any dollar is the equivalent of any other dollar. It should not matter to a bitcoin or dollar holder which particular dollar or bitcoin they get. NFTs, however, can be distinct; no one is like the other, unless the creator of the NFTs has designed for a certain number of NFTs to be the same. A good analogy to understand the difference between a fungible token and an NFT is in the pickup line from kindergarten. When you pick up your kid, does it need to be *your* kid, or could it be *any* kid? It probably needs to be *your* kid, clearly distinct and differentiated from all the others.

An NFT can be linked to a piece of content, like a JPEG of a cartoon cat, a music file, photo, video, or document. It can be a badge, certificate, or medallion that entitles the holder to other privileges, such as earning rewards, having eligibility to receive other subsequent tokens or NFTs, obtaining authorization to contribute to a project, or enjoying the right to get paid for doing so. An NFT can represent membership to a community. Chat platforms such as Discord enable users to connect their Web3 wallets such as MetaMask to prove they hold a given NFT, and once proven, admit them to exclusive online communities, the equivalent of Soho House for Web3. NFTs can represent ownership of real-world or digital goods, such as a Web3 version of a receipt or certificate of authenticity. They can represent a land title, an individual's subscription agreement to be a member of a DAO, a software license, or just about anything they are programmed to do. Similar to fungible tokens, NFTs can be distributed in a variety of different ways: for purchase on a website, marketplace, or peer-to-peer, via airdrop or claimable, or by retroactive distribution to reward past behavior. Like other types of tokens and Web3 structures, not all NFT designs and distribution mechanisms are necessarily legal in every jurisdiction around the world, and they may become subject to emerging legal and regulatory frameworks.

Though NFTs can be distributed to wallets on a blockchain network in many similar ways to fungible tokens, they serve a distinct purpose on the Web3 Internet of Value. Fungible tokens such as bitcoin and ether can play the role of digital money—that is, a medium of exchange that facilitates the pricing and exchange of other types of value—but NFTs are digital objects.

An analogy for thinking of them is as the internet-native version of the products that a consumer would buy using traditional money. There are far more use cases for NFTs than Twitter profile pictures (PFPs) to signal to followers a user is Web3-savvy or has the resources and taste to buy a luxury item, though the social media PFP use case surely should not be ignored. Imagine that a person exchanges dollars for a pair of Nike sneakers. In Web3, that person might exchange ether for an NFT of Nike sneaker wearables for their avatar in the metaverse (a concept we'll explore in greater depth in Chapter 4, but that might quickly be described for now as an immersive virtual world that integrates physical and digital realities). Beyond the obvious differences between physical and digital objects, the NFT can have capabilities that the physical sneakers do not, thanks to the fact that they live on a decentralized blockchain. They do not necessarily lose value from "use" like a physical sneaker. A liquid market for the Nike sneaker NFTs may exist such that the buyer could at any moment frictionlessly resell the sneakers to another buyer at the same or a higher price. There are certainly some rare physical Nike sneakers with significant resale markets, but these are a low percentage of all Nike sneakers, and even if a worn pair of sneakers can be resold, resellers typically charge high fees for intermediating the transaction. Most worn Nike sneakers could only be sold for a tiny fraction of the original price in secondhand clothing stores.

Thus, NFT buyers, as long as they believe they are buying NFTs that other buyers will want in the future, can buy with the mentality that they are investing, or at least parking funds, in the NFT asset, as opposed to consuming the funds, which is generally the buyer's mentality when purchasing a regular pair of Nike sneakers to wear. Buyers with this mentality can be comfortable spending more on Nike sneaker NFTs than on regular Nike sneakers.

The owner of a Nike sneaker NFT can also *fractionalize* that NFT into tiny pieces that are each fungible tokens—perhaps as a way of involving a community in ownership, perhaps simply as a matter of unlocking liquidity, perhaps as a bit of both. Sometimes fractionalized NFTs are used in DeFi liquidity pools, such that a user can generate LP tokens and claim rewards for provisioning liquidity in that form. In this case, the sum of the parts can be greater than the whole; the value of the total fungible tokens that make up a single fractionalized NFT can be higher than the value of the single NFT. This has been the case for several NFT collections—notably CryptoPunks, one of the first NFT series and the series that has held the highest market capitalization of any collection over a long period, individual NFTs from which have been fractionalized and yield farmed in some of the largest liquidity pools. (*Yield farming* is the activity of getting rewards for providing liquidity to a DeFi protocol, sometimes re-compounding that liquidity in multiple layers across multiple protocols to enhance rewards, a practice that can yield a whopping annual percentage yield, or APY, but can also be dangerous and collapse. More on this in the DeFi section.)

Fractionalization is far from the only use case for NFTs in DeFi. Tools and platforms are emerging that allow NFT holders to borrow against their NFTs by using them as collateral. Though art is only one NFT use case, it offers a useful analogy to this scenario. Many collectors who buy expensive art from large galleries or auction houses like Sotheby's and Christie's take out loans against their collections, giving them liquidity—perhaps for further investments they believe will pay dividends. If the art is considered stable enough in price, such as a Monet or Picasso, lenders are willing to offer them loans collateralized by their art. Sotheby's has its own division called Sotheby's Financial Services that offers these loans directly to Sotheby's art buyers, leveraging their expertise in evaluating the market risks associated with art assets to price the interest of the loans.

A holder of an NFT in a blue-chip collection—that is, in one of the most trusted collections that have been relatively stable in price over time, such as CryptoPunks or Bored Apes—could similarly access tokenized liquidity from collateralizing a loan with that NFT. This could become a profitable exercise for both borrowers and lenders, as long as NFT lending services or dapps manage to price the risk accurately. When a user buys an NFT, especially one with an established price history that is appealing to lenders, far from consuming the tokenized funds they used to buy it, they potentially unlock a new source of value capture. Thanks to tokenized rewards and yields from DeFi, a Nike sneaker NFT on a decentralized blockchain can be preferable to buyers over a physical Nike sneaker.

Producing sneaker NFTs, to extend our example, is also good for Nike. Physical sneakers require up-front costs from manufacture through every step along the supply chain, from factory to wholesaler to retail point of sale. Factories and distributors are intermediaries that charge fees to Nike all along the sneaker's journey. Each step along the physical supply chain introduces risk of loss, theft, and damage. Unlike a physical sneaker, no significant up-front costs are required to produce an NFT, except for those associated with graphic design and any 3D or metaverse capabilities programmed into the NFT. Nike can be more creative with its sneaker designs, free from the limitations of physical materials, utility as shoes, or the laws of gravity. As long as the NFT wearable adheres to the standards of a Web3-enabled metaverse so an avatar can wear them, there's no reason why the sneakers can't be made of rainbows, on fire, or allow an avatar to walk on the ceiling.

Nike could also be motivated to produce NFTs for business reasons. People will continue buying physical sneakers, and Nike will continue producing them. By starting to sell NFTs, Nike can introduce a new, noncannibalistic product line with a low ratio of production cost to sale price. There's no reason selling NFTs would cause people to buy fewer physical sneakers, but every reason to start selling a new, high-margin type of item, with instant global distribution.

Skeptics often wonder why utility buyers as opposed to speculative investors would purchase a Nike sneaker NFT. All the digital shoes in the world wouldn't prevent someone from going barefoot. But these skeptics probably don't know any gamers or young children. The kids in my family always surprise me at holiday time; they much prefer in-game assets on Fortnite or Roblox over physical presents. But it really shouldn't be all that surprising—it's only natural, after all, that some of the objects we desire the most are native to the digital environments we inhabit with increasing frequency. Long before Web3 and since the early days of the web, gamers have paid hefty price tags for the most desirable virtual goods.[3] They are often the early adopters that point to larger emerging trends. Fashion, art, music, and more are taking up residence in metaverse worlds and producing NFTs in Web3-enabled metaverses.

Indeed, creative industries—especially ones such as music whose revenue streams have been severely affected by Web2—are among those most intrigued by the possibilities of Web3 in general and NFTs in particular for capturing value. When designing a new NFT collection, for example, creators can choose to program in royalties that enable them to collect a percentage of sale value every time an NFT is resold after the original sale. This enables creators to continue benefiting indefinitely from the success of their work, in contrast to the traditional and Web2 paradigms in which intermediaries pay artists fees and then go on to reap the lion's share of benefits over the lifetime of the creative work. Though not all NFT collections are designed to include royalties for original creators, and some may include this or royalties to other parties (like the software company that created the API used to mint the NFTs), the potential for including this value-add to creators is part of what makes Web3 preferable over other creative platforms.

The growing popularity of digital goods is a poignant example of the world of bits overtaking the world of atoms. However, NFTs need not be divorced from the physical world. Marketers have termed these products *phygital* that span the physical and the digital. A retailer such as Nike could sell sneaker NFTs that allow holders to claim a pair of physical shoes, online or in a store. Nike could reward past buyers of physical sneakers with look-alike sneaker NFTs that could be collected and traded, or function as wearables for avatars in metaverse worlds. NFTs could simply accompany any physical shoe purchase and function like a more credible version of a receipt, reducing friction in the resale market with provable authenticity and a traceable chain of ownership.

Nike also sells extremely rare, high-priced sneakers. The rapper Drake bought a pair of solid gold OVO x Air Jordans (which, though shaped like sneakers, cannot be used as shoes any more than a digital sneaker) for over $2 million in 2016. Most rare sneaker buyers are collectors and resellers, and some may prefer not to store the sneakers themselves, where they could be subject to damage.

An NFT representing ownership of a rare sneaker could be traded between collectors with the option to take physical delivery of the item until a collector decided to do so. With an NFT, Drake could safeguard his golden sneakers from loss or damage by clumsy or sticky-fingered house guests.

As more and different types of businesses begin producing NFTs, they will continue to innovate novel ways to link physical and digital goods into the phygital. They are already doing so in droves. Bulgari recently broke the record for the thinnest-ever watch with the Octo Finissimo Ultra, only 1.8 millimeters thick, which it sold accompanied by an NFT that attests its authenticity. Pace Gallery, one of the world's leading art galleries, launched an NFT collection called "Moon Phases" with artist Jeff Koons, which involves sending miniature sculptures of the moon literally to the moon itself, on a lunar lander. Earthbound sculptures corresponding to the miniature ones landed on the moon will then be sold with accompanying NFTs. Outside of art and business, governments such as Dubai are already registering land titles to the blockchain, unlocking efficiencies for administrators by moving authoritative documents on-chain.

While the world of bits is still innovating more and faster than the world of atoms, perhaps its momentum will drive progress in the latter. The blockchain has no hand with which to reach out and grab onto the physical world, but through incentives, such as the incentive for an owner to keep a valuable object with its associated NFT, and strategically placed QR codes for claiming NFTs in the physical world, the two can be intertwined.

NFTs can represent a modestly priced pair of digital sneakers or immensely valuable digital and physical objects, such as Bulgari's thinnest watch or CryptoPunk #3100, which sold in 2021 for a whopping $7.58 million. Across the price spectrum, scarcity is key to determining the value of an NFT. In the physical world, it's crystal clear how scarcity works: for example, Bulgari produces only a limited run of the Octo Finissimo Ultra. In a free market, price is determined by the ratio of demand to supply, buy pressure to sell pressure. Stimulating demand while limiting supply is a formula for driving higher prices. Thanks to the fact that NFTs live on decentralized blockchains with their ingenious solution to the double-spend problem, creators of NFTs can easily limit supply in the digital world and make their NFTs rarer.

Critics of NFTs who don't understand the technology wonder why buyers would pay so much for a JPEG when they could just "right-click save" to their computer and copy the image an arbitrary number of times. If this is how NFTs worked, the supply would be potentially limitless, and regardless of any demand, prices would be driven down to zero.

Instead, each NFT is a unique token on the blockchain. As each new block is added to the chain, the ledger updates, immutably recording all NFT transactions, the public addresses of the wallets that own NFTs, and which NFTs they hold, just as it does with fungible tokens like bitcoin and ether. Although anyone online can right-click save and copy a JPEG associated with an NFT, that doesn't actually transfer ownership of the NFT on the blockchain. A JPEG of a Bored Ape or CryptoPunk has no resale or status value. There is no DAO or pool of liquidity demanding that JPEG without the proof of authenticity that comes from the NFT's history on the blockchain. Merely copying a Bored Ape image as one's Twitter avatar has no cachet, since anyone could do the same; what does have cachet is the special hexagonal border Twitter puts around the same Bored Ape image, indicating that one has connected one's wallet and Twitter has verified ownership of the NFT on the blockchain: it's the authentic digital object. So although an image associated with an NFT is easy to copy and save, and there may be some personal or artistic value to using such an image, the image per se has no token market value— unless one is trying to scam the uninitiated on NFT marketplaces like OpenSea, which are increasingly cracking down on counterfeits.

When buying an NFT, it's essential to verify that its chain of ownership traces back to the correct original producer. Fortunately, this is trivial in the UX of OpenSea and other marketplaces, and fairly straightforward on-chain using a block explorer such as EtherScan. It's yet another advantage NFTs have over physical goods. Although it's sometimes difficult to verify whether a Prada handbag is real or fake, Prada—which has launched its own NFTs in partnership with Adidas—can easily trace a Prada NFT back to the wallet that conducted the original Prada mint or to an unlucky fraudster. So can its buyers. Every year luxury brands lose billions of dollars to counterfeits. NFTs elegantly solve this problem and make it easier than ever to verify the authenticity of any digital item.

So many of the benefits of NFTs derive from the fact that they sit on a decentralized blockchain. However, because NFTs have become more widely adopted and the subject of intense (not to say sensational) media interest, there are a few caveats to keep in mind. For one thing, some opportunistic creators have taken to launching digital goods that they call *NFTs*, but that are not actually on a blockchain. For another, it matters a great deal which blockchain is used to create a given NFT; not all blockchains confer the same benefits. Generally speaking, Ethereum and other Ethereum-compatible (EVM-compatible) blockchains are preferable over other, newer blockchains for NFT projects. For one thing, smaller, newer platforms may shut down—after all, it's commonly said that 90% of startups fail—and when they do, the tokens and value on their networks can disappear.[4]

Another significant advantage to the Ethereum and EVM-compatible blockchains is the proliferation of DeFi tools and platforms, as well as DAOs such as Flamingo with large pools of tokenized capital in ether and other ERC-20 tokens that put demand pressure on NFTs, allowing owners to achieve liquidity and fluidly exit NFT positions. (It's possible that this may change in the future as other networks like Solana and Flow promise lower or zero gas fees that make it more sensible to conduct more transactions with lower-price NFTs. Users, however, are probably well advised to make sure their NFTs live on an actually decentralized blockchain that is not likely to shut down or disappear, which would mean losing access to their assets.

On a basic architectural level, other Layer 1 platforms—that is, base blockchain networks—are more centralized (or less decentralized) than Ethereum. In designing a blockchain network, there are inherent trade-offs in optimizing for security, scalability, or decentralization; this is often referred to, originally by Vitalik Buterin, as the *blockchain trilemma*. From the beginning, Ethereum optimized for decentralization to serve as the base settlement layer (Layer 1) for a complete alternative financial system. A network may optimize instead for scalability—that is, an ability to handle more transaction throughput, cheaper and faster—at the cost of less decentralization, but such a network bears a greater risk of being shut down or manipulated by a small group of centralized actors. The less decentralized the platform, the more vulnerable it is to being frozen, going offline, or shutting down, whether by accident, "for maintenance," or under pressure from authorities. During an offline period, NFT holders might not be able to access their wallets or the value they thought they were holding. Wallets that can freeze or be shut down are fundamentally not self-sovereign. "Not your keys, not your crypto" is a popular expression in the space, because unless you own the private keys to access your assets on-chain, someone else could arbitrarily decide to remove or destroy them.

That said, cheaper and faster transactions and greater scalability are certainly desirable, and another significant draw of newer, more scalable networks has been less environmental impact through energy use. Both Bitcoin and Ethereum blockchains began by using proof-of-work mining algorithms, which require large amounts of electricity to power. On September 15, 2022, the Ethereum blockchain successfully navigated a long-planned transition to a *proof-of-stake* consensus algorithm, a massive feat of worldwide coordination called *the Merge*. In proof-of-stake, instead of computers competing to cryptographically secure the network by deploying hashpower to solve resource-intensive hard math problems, each computer can stake an amount of ether that it forfeits if it tries to incorrectly portray the new state of the ledger. Thus proof-of-stake theoretically can achieve the same level of cryptographic security as proof-of-work while remaining decentralized, lowering the total amount of computing power Ethereum requires. It is estimated that the Merge has decreased Ethereum's energy needs by over 99% and possibly worldwide electricity consumption by as much as 0.2%.[5] Sustainability advocates that have continued to express concern about NFTs on the Ethereum blockchain even after the Merge are probably unaware that the transition to proof-of-stake has already solved their problem.

Although the Merge successfully decreased Ethereum's energy needs and introduced other benefits, moving to proof-of-stake does not meaningfully improve the network's gas costs or scalability. Improvements such as sharding—another long-planned improvement to Ethereum that allows many transactions to be executed simultaneously—are expected to help address some of these needs.

A nearer-term option for decreasing costs and increasing scalability for NFT transactions is layering a second blockchain, a Layer 2, on top of Ethereum. Imagine a casino where a gambler buys a stack of chips at the beginning of the night, plays 100 games of blackjack, and then settles out into dollars before going home. The gambler wouldn't want to settle out after each game; this would be unnecessarily time-consuming. By this analogy, it's unnecessary in the course of Web3 gameplay or a trading session to settle out to the Ethereum Layer 1 blockchain with each transaction. A participant could play their game or conduct their trades on a more scalable Layer 2 on top of Ethereum for low or zero gas fees, and then settle out, incurring gas costs, to the relatively slower Layer 1 at the end of the session.

The most popular mechanisms for Layer 2 scaling solutions on Ethereum are sidechains and rollups. Of the leading Layer 2 solutions, Optimism and StarkNet are rollups, and Polygon is a sidechain, though Polygon Hermez in the Polygon Suite is a rollup. In a *rollup*, the computation of transactions happens off of the Ethereum blockchain (off-chain), but the representations or proofs marking the correct state of the ledger are marked on-chain, so that whenever the user settles out to the Ethereum blockchain, using our casino example, the end state is always accurate. This happens once tokens are bridged back to the Layer 1 from the Layer 2, as long as the Layer 2 correctly executed the transaction. By contrast, a *sidechain* is a separate blockchain network running in parallel to the Ethereum blockchain, so whenever the user wants to settle out, they can move their assets back to the main Layer 1 chain. These are both effective mechanisms for adding scalability on top of the Layer 1 Ethereum blockchain that enable the user to settle out whenever they please.

My opinion is that Ethereum will continue as the platform of choice for NFTs and dapps, sometimes paired with Layer 2 solutions. Ethereum has prioritized decentralization, making it the best base settlement layer. As the first mover in the programmable blockchain space, it has gained the network effects of having the most users, the largest liquidity pools (such as DAOs) for applying demand pressure, and the most diverse and advanced DeFi tools. And it plans to increase scalability since transitioning to proof-of-stake through mechanisms such as sharding. Meanwhile, an increasing number of NFTs, especially lower-priced and gaming NFTs, will be built on compatible Layer 2 platforms like Polygon, which is already becoming popular for this purpose.

Others have different opinions. As we say in Web3, *dyor* (do your own research). And beware: often the people hawking other Layer 1 blockchains (in Web3 speak, *shilling*), as well as fungible tokens and

NFTs, are investors in those tokens simply trying to "pump their bags"—that is, to convince others their position is correct, motivated by personal financial gain instead of true conviction. In the interest of transparency, let me disclose that I am a holder of ether, bitcoin, and a host of other fungible tokens with technology, capabilities, and teams I believe in. But I'm also a person writing a book in earnest about Web3 that I hope will serve its readers, so perhaps you can trust me. Please still dyor.

Decentralized Finance (DeFi)

Arguably, Bitcoin was the first DeFi application. *Decentralized finance* is exactly what it seems: finance on a decentralized blockchain, instead of through a centralized entity like a bank. Bitcoin was the first successful implementation of the idea that it was possible to construct a secure digital money system without involving national governments or banking intermediaries. As discussed, its design left Bitcoin fairly limited to the purpose of serving as money. Ethereum would innovate beyond the original Bitcoin concept to serve as the foundation for an alternative financial system, also undergirded by secure digital money but additionally capable of more than sending money from wallet to wallet around a network.

The traditional modern financial system (TradFi) isn't simply a network of accounts sending money to each other. It is a sophisticated web of financial tools and products. Securities, bonds, commodities, real estate, and currencies are the staples of TradFi. All manner of speculative products are layered on top of them for investors to borrow, lend, and bet on their performance, from mortgages for real estate to futures and derivatives. This industry is almost completely digital: 92% of the world's money exists in digital form only.[6]

What is remarkable about DeFi is not that it digitizes the global financial system, which already from the user's perspective appears to have happened in Web2—though in fact large swaths of the process of buying and selling traditional financial products are still written on pieces of paper, even if they are presented digitally in the user interface of a trading account such as Charles Schwab. Rather, it's that DeFi has built an alternative financial system that lives alongside the traditional one, giving users the option to avoid intermediation, have custody of their own assets, and access novel financial applications only available thanks to Web3 capabilities.

Rome wasn't built in a day, and today crypto comprises only about 1% of the $96.1 trillion global GDP.[7] However, DeFi is growing rapidly in terms of wallets created and number and volume of transactions. As cryptocurrency becomes a larger portion of global financial activity, DeFi tools and applications will continue to grow, and as they do, earn battle scars from bouts with hackers and regulators. These will spell doom for some DeFi protocols and make others stronger. The most fundamental DeFi applications allow users to borrow and lend with their tokens.

Compound, Aave, and Maker are some of the best-known platforms for DeFi borrowing and lending. Similar to TradFi borrowing and lending facilities, users can deposit a given token into a shared pool and receive interest, or deposit tokens as collateral and borrow a smaller number of tokens and pay interest. Unlike TradFi, borrowers need not submit application forms nor disclose FICO scores. No one is rejected, because users submit the collateral on-chain in advance of receiving the loan. The entire process executes automatically and transparently on the blockchain. These are dapps built on smart contracts, not discretionary processes with rates determined behind closed doors by groups of people.

DeFi lending sites openly publish on their websites the annual percentage yield (APY) they are offering for particular tokens, along with their other key metrics, such as total value locked (TVL)—that is, the current value of all of a DeFi protocol's crypto deposits. APY lets users know the returns they can expect, and a high TVL offers assurance that others are using the platform—DeFi platforms compete to draw the most users and earn the highest TVL. Newer DeFi protocols with low TVL typically haven't passed as many code audits or battle tests by hackers, so users may be wary of engaging. However, newer platforms sometimes offer higher APY than established ones, so in choosing a platform, DeFi users may balance platform risk against optimizing yield. Unlike reversing a bank transaction, when DeFi hacks happen, there's often no recourse for users to recover their funds. That's the price of decentralization.

To many, the benefits are well worth the risks. Compound, Aave, and Maker adapt TradFi borrowing and lending to a native Web3 environment; though they are run by different financial mechanisms, they have several key benefits in common thanks to Web3. First, they aim to be noncustodial, which means at no point in the transaction could the Aave team, for example, hold the pool of tokens in wallets that it owns. Ideally, the capital in the Aave protocol could not be frozen at some authority's demand ·or withdrawn by its employees. Second, they are permissionless; anyone with a wallet and tokens can participate. Third, they are trustless. There is no need to trust any individuals working on Aave. The Aave protocol is open source code that anyone with the right skill set can inspect and diligence. Finally, although banks encumber users with loan application processes that include heavy KYC/AML (know your customer and anti–money laundering) requirements, DeFi platforms originate loans with the press of a button. This also protects user privacy. The DeFi protocol can only see public wallet addresses; users are effectively pseudo‑nymous and need not disclose identity information.

As DeFi projects abandon centralized entities in favor of governance by their stakeholders in DAOs, they simply become pieces of self-executing software that exist online. Even if a regulator wanted to shut down such a protocol, they theoretically could not, because it is maintained by a decentralized network of computers, as opposed to a single entity or individual. Over the next decade, pressure to regulate DeFi is likely to increase. This will cause a bifurcation between DeFi that is maximally decentralized from top to bottom, where no entity or individual could discretionarily change the code base or turn it on or off, and those with centralized points of failure. Protocols with centralized elements may be pressured to introduce KYC/AML processes to identify users, destroying the benefits of privacy and permissionlessness, while increasing costs and loan origination time. Fully decentralized DeFi protocols can avoid these pitfalls by using Ethereum to decentralize the contracts, a solution like IPFS (the InterPlanetary File System) to decentralize the files and data, and pseudonymity by public wallet address to conceal their personal identity information.[8]

Unsurprisingly, there is a trend among DeFi protocol founders to operate as *anons*, or pseudonymous accounts, often with whimsical names. After all, if the code base is open source for anyone to diligence, there is no need for a trusted individual at the helm of the project—or is there? Although privacy and pseudonymity, especially concerning money transactions, have been important in the Web3 community since Satoshi

Nakamoto (and in the hacker and cryptography communities that gave rise to it for far longer), project leaders nonetheless bear a greater share of responsibility than the average user. We will shortly consider one of the most (in)famous anons, SushiSwap founder Chef Nomi.

To understand SushiSwap, though, we must first introduce the DEX. As mentioned previously, a DEX is a *decentralized exchange*. It serves the same core purpose as a centralized exchange, like the New York Stock Exchange, which enables people to buy and sell stocks. The difference is that the NYSE intermediates each transaction: the NYSE keeps track of the trades, using custodians to hold the stocks being exchanged and charging a fee for providing intermediation. DEXes handle tokens instead of stocks and use automated market makers (AMMs) to directly match supply and demand for tokens. Each transaction is peer-to-peer, such that the DEX never custodies the tokens over the course of the transaction, with tokens locked in a smart contract from which users can withdraw them.

Not every crypto exchange is a DEX. Many of the best-known crypto exchanges, such as Coinbase, Binance, and the now-defunct FTX, are far from being DEXes. Structurally, they are traditional exchanges like the NYSE that happen to deal in tokens rather than stocks. They custody the tokens themselves, instead of users holding them in self-sovereign wallets. This leaves users exposed to the pitfalls of centralization; their accounts can be frozen or shut down, either in error or if the company faces external pressure. The industry has a term for these centralized exchanges (CEXes), as well as centralized platforms for borrowing and lending tokens like BlockFi and Celsius (both well known but now collapsed): *CeFi*, or centralized finance, closer to TradFi in many ways than DeFi. They may offer users exposure to token investments or yields, but they obviate any user benefits derived from Web3. Their support lines are famously teeming with customers whose accounts have been arbitrarily locked or frozen. Conversely, it can be considered a benefit of CeFi companies that they have the power to reverse erroneous transactions. Their other main benefit is simpler UX for new users who arrive without self-sovereign wallets like MetaMask. As new users enter Web3, they often buy their first tokens on a CEX, then as they learn more, move their tokens to a MetaMask wallet, then start playing with NFTs or DeFi.

A tsunami of these fresh users poured into DeFi during summer 2020. This was the first summer of the global COVID-19 pandemic, and people all over the world were quarantined in their homes where, free from the distractions of family barbecues or trips to the seashore, they found themselves with time to spare for exploring some rapidly transforming digital realms. The crypto winter that hit in 2018 and sunk to frightening depths in 2019 had begun to defrost, and prices for tokens like bitcoin and ether had begun their tenuous path to recovery. It was, perhaps, crypto spring. Marooned in their coding chairs, Web3 devs revved up their engines of innovation.

It was probably either Kain Warwick, founder of the DeFi derivatives trading platform Synthetix, or Andre Cronje, the developer of Yearn Finance (YFI), who first came up with yield farming; in true Web3 style, each has attributed the invention to the other. This was simply the idea that new DeFi protocols could attract liquidity by offering rewards for staking particular tokens. The protocols would deploy the liquidity on their own platforms or, in the case of YFI, other DeFi lending platforms, to earn the yield they would distribute to their users as rewards. In exchange for providing liquidity to one of these pools, a user would receive LP (liquidity provider) tokens representing their claim to the assets in the pool. These could have governance functions or entitle holders to separate governance tokens, enabling them to vote on decisions such as where the liquidity pool would be deployed to generate yield and how to distribute yield to LP token holders.

Liquidity pools sprung up all over Web3. Famously, many were named after foods. The summer of 2020 was affectionately nicknamed "DeFi summer" or "DeFi foodcoin summer." APYs could sometimes surge to over 1,000% on platforms such as Pickle Finance and Yam Finance. A daring yield farmer could use the rewards from one pool, or *farm*, and use those rewards to deploy liquidity to earn LP tokens and thereby rewards in a third farm, and so on, ad infinitum. Tokenized fortunes were made. The new crop of superstar developers and hot DeFi projects—many, such as Based Finance, with anon founders—came to dominate the conversation on Crypto Twitter. Their Discord servers overflowed with yield farmers seeking alpha (in finance speak, looking for information that leads to excess returns), trading memes, meeting fellow DeFi enthusiasts, and starting new experimental projects. During the isolating pandemic, these communities were oases for social life where the memes flowed freely.

Fortunes also vanished as quickly as they were made. Yam collapsed due to a flaw in a single line of code. Like so many foodcoins, it had launched without substantive code audits. Grasping at the promise of high APY, yield farmers had poured into hastily built liquidity pools, ignoring the warnings signs on their websites disclaiming the code hadn't been audited and encouraging users to dyor. Harvest Finance (its name a play on yield farming) suffered the worst hack of 2020 and lost $34 million in user funds.[9] Exploits of DeFi protocols would claim a whopping $2.2 billion in 2021.[10]

Yield farming was a dangerous game during DeFi foodcoin summer, but it had massive salutary effects for the overall health of the Web3 ecosystem. For one, almost all the lightning-speed exchanges of tokens took place on DEXes, thanks to the fact that they are permissionless, unlike CEXes. There was so much volume that leading DEXes such as Uniswap processed more transactions than the top CEXes such as Coinbase, which were previously known to process the most volume.

DeFi was overtaking CeFi, a victory for the decentralization evangelists of Web3. DEXes started offering "boosted" rewards for providing liquidity to their liquidity pools, taking advantage of the yield farming trend, to help arm their AMMs with a diverse supply of tokens. More DEX use led to a better experience for users, who could find almost any token they wanted on a DEX, and for Web3 projects with tokens, which could find distribution for their token on a DEX even if it wasn't listed on any CEX. The process for getting listed on a CEX was notoriously slow, opaque, and labor-intensive. Sometimes CEXes would demand high fees from projects to list their tokens, knowing they had a lock on distribution. The competition from DEXes disintermediated the CEXes, giving token projects clamoring for listings another, easier option. Marketers for token projects rushed to set up liquidity pools for their tokens on DEXes to stimulate demand. Tokens started launching directly on DEXes, and the initial DEX offering (IDO) was born.

But the DEXes were not immune from danger. Similar to DeFi protocols, hackers target them constantly. And during DeFi summer, the competition between DEXes for volume reached dramatic heights. Now we return to the pseudonymous Chef Nomi, who forked the Uniswap protocol and launched SushiSwap, initially a clone of

Uniswap with different branding that later added different features. SushiSwap launched with a *vampire attack* on Uniswap, enabling anyone with Uniswap LP tokens to stake them in SushiSwap and receive its associated tokens called SUSHI. SushiSwap planned to transform the Uniswap LP tokens to SushiSwap LP tokens, which would essentially steal all the liquidity from Uniswap. The attack wasn't as successful as planned, but from it the DeFi community learned that "degenerate yield farmers"—or *degens*, for short; *degenerate* in the gambling sense of betting significant sums without due diligence, a pejorative that many embraced as a badge of risk-taking honor—had little loyalty to one DEX over another, especially if it used a similar, tried-and-true code base and offered a chance to capture value, even by participating in a hostile maneuver.

The founders of the two DEXes battled on Crypto Twitter. Ultimately, Uniswap would prevail, continuing to flourish along with other popular DEXes such as Curve and Balancer. Chef Nomi would go down in disgrace, not for essentially copying Uniswap's code, but for a far worse transgression in the eyes of Web3. Toward the end of summer 2020, Chef Nomi cashed out $14 million of ether from the SushiSwap developer fund wallet to their personal wallet, which crashed the price of SUSHI as frightened degens pulled their tokens out of the DEX. Their concern was that Chef Nomi had perpetrated a "rugpull," the cardinal sin for DeFi founders: after growing volume or token value on a protocol, they suddenly allocate the tokens (typically from the protocol itself, but in this case, from a wallet of funds to be allocated to protocol developers) to their personal wallet and disappear.

SushiSwap continued to operate, but to restore trust in the platform, replaced its anon founder with a team mostly composed of people using their real names instead of pseudonyms. Some degens briefly learned their lesson. Protocols learned the benefits of project teams using their real names. Others continue to operate as anons. Chef Nomi returned all the ether to the SushiSwap treasury six days after withdrawing it. They tweeted an apology and promised to allow Sushi's community to decide if they deserved payment as its founder. The community had bigger problems than how to reward their prodigal founder, however, and voted to use all of the returned ether to buy SUSHI in an attempt to prop up its cratering price. That buyback was not immediately successful, but Sushi would nonetheless live to exchange another day. As of this writing, Chef Nomi has not been heard from since the day after their apology.

Given the aforementioned scandals, hacks, and volatility, it may—or indeed, may not—be surprising that one of the more important DeFi innovations is something called a *stablecoin*, which is a token, often an ERC-20, whose value is pegged to the price of a fiat currency (often the US dollar). The health of DeFi depends on the availability of well-constructed stablecoins. When people buy an asset, they typically buy it using a currency that does not massively fluctuate in value. The volatility of floating crypto assets such as bitcoin and ether can therefore make them undesirable for certain types of transactions and DeFi activities. That stablecoins are, well, stable relative to fiat currency makes them essential to many DeFi products; they also allow users to hold tokenized representations of fiat currency on-chain instead of paying fees to CEXes to convert out of crypto into fiat.

Some stablecoin attempts have failed disastrously (such as the Terra UST stablecoin and its floating token, LUNA), but this is due to ill-designed tokenomics, not any flaw in the inherent concept of stablecoins. USDT, which is produced by Tether, and USD Coin (USDC), produced by a consortium led by Circle, are the most widely used centralized stablecoins. Both are asset-backed stablecoins controlled by private companies. Dai, an offering from MakerDAO, is the most used decentralized stablecoin. These organizations take different approaches to producing stablecoins, but as of this writing have all maintained success at keeping the value of one coin equivalent to one US dollar.

4

The Metaverse

THE SCIENCE FICTION author Neal Stephenson coined the term *metaverse* in the novel *Snow Crash*, 30 years before Facebook would rename itself Meta and declare its intention to occupy a digital realm of the same name. *Snow Crash* tells the story of Hiro, a pizza delivery boy in an anarcho-capitalist future. A hacker in his spare time, Hiro draws a cast of computer nerds, tycoons, and mafiosos into daredevil schemes in an immersive, virtual reality, massive online multiplayer game called the Metaverse. Before nearly anyone else, Stephenson imagined that people would want to merge their physical and digital realities and dive deeper into inhabiting virtual worlds. He turned out to be right. Science fiction classics like *Snow Crash* and Ernest Cline's 2011 *Ready Player One* make the natural leap from contemporary capitalism to imagining companies in digital battles to control and monetize metaverse worlds. The bad guys will even sacrifice the quality of the game, or make it too expensive for most people to play, in the name of short-term profit. Nolan Sorrento, the villain of *Ready Player One*, runs a company called Innovative Online Industries (IOI). IOI is determined to control the open, accessible OASIS, an immersive gaming world that is a refuge for young people away from their dystopian physical reality, to privatize it behind a paywall and charge high fees for access.

When a company like Facebook creates a metaverse world—a company that already makes users into its product, while siphoning the value of user-created content to its shareholders—the Web3 community knows a real-life IOI when it sees one. Comparisons between Mark Zuckerberg and Nolan Sorrento only got louder after Meta's April 2022 announcement of a 47.5% fee for creators to build in their world.[1] Mostly, though, Web3 is focused on building its own metaverse worlds—and there is significant variation and ongoing evolution of what, exactly, that means.

Unlike a Web2 game or virtual reality experience, Web3-enabled metaverses are open environments for creating content and value using fungible tokens, NFTs, and even DeFi. A set of standards common to Web3 metaverse worlds enables assets to be portable between metaverses. Though large and powerful companies are taking aim at "the metaverse" as their next territory to conquer, it can be hard for incumbents who are set in their ways to pivot their business models and tackle new challenges. The field of traditional and Web2 companies attempting to enter Web3 is littered with the bodies of failed ventures. Facebook's own most recent attempts at building in Web3, the Libra and Calibra projects, were notorious failures. In an arena where incumbents don't always win and smaller, more agile players enjoy advantages, Web3 metaverses can prevail just like Wade Watts, the young protagonist of *Ready Player One*, who triumphed over Nolan Sorrento and IOI. Web3 has a fighting chance to become the economic foundation for the metaverse worlds where we will choose to live.

In the United States, where people spend an average of six hours per day accessing digital media, "real life" is arguably already a mix of digital and physical life.[2] Virtual reality is becoming increasingly immersive, not only as a place to play games, but as the office where we work, the supermarket where we shop, the mall where we try on clothes and shop for furniture, the venue where we attend a concert, the stadium where we watch the big game, the gym where we exercise, the quiet space where we meditate, the cafes and restaurants where we socialize with friends, the place where we visit our lawyer, doctor, therapist, notary, or priest. In theory, we could spend all our waking hours in a virtual reality world or metaverse (and our sleeping hours collecting biofeedback data for our avatars the next day), especially with the broad definition of being "in the metaverse" that I personally prefer, where we're in the metaverse every time our attention is focused on or mediated by a digital environment. This includes when we are looking at our computers or phones—these are simply less advanced and immersive metaverse worlds. It's likely that, in the coming years, we'll move back and forth between more and less immersive worlds, toggling between using traditional devices (such as smartphones) and engaging in immersion delivered via wearables (headsets, glasses, haptic suits, contact lenses) or, perhaps, implants in body and brain—the virtual, digital world merging with our physical reality and vice versa.

When we spend time on Web2 platforms, our data and identity information are being monetized to sell to advertisers, but for most of us, the vast majority of our daily lives still take place off Web2 social media (even if constantly intersecting with and being affected by it). Most of us wouldn't stand for a Web2 business model that extracts our data for profit and feeds us behavior-altering advertisements in the place where we actually live most of the day. Instinctively, that would feel like a dystopia. And users, especially in the most developed post-industrial markets of early adopters, are wising up to the Web2 business model and increasingly rejecting it. This was demonstrated recently when Apple introduced an option for iPhone users to prevent apps (like Facebook) from collecting certain personal data. When presented with this option, enabled by easy UX at the push of a button, so many users chose privacy that Meta estimates the change will cost it $10 billion in lost 2022 advertising revenue, the equivalent of 8% of Meta's ad revenue the previous year (and recall that advertising accounted for 97% of Meta's total 2021 revenue).[3]

A decade ago privacy advocates were laughed out of the room when they claimed that users would insist on personal data privacy. Today the idea that Web3 will provide the economic foundation of the metaverse where we live seems ludicrous to skeptics, who see companies like Meta aggressively pushing to build in the metaverse and assume the incumbents always win, forgetting that public preferences can change and quite quickly, catalyzed by the right UX. User preferences will matter in the metaverse, and if we are to live there—or if we already do—there's every reason for users to demand Web3.

With the right economic foundation, the metaverse sounds more like a utopia. Creators and artists can stretch their imaginations to the limit without the constraints of gravity or supply chains, escaping from the linear materials economy. If they offer their work to others in the form of NFTs, they can sell their work with a low barrier to entry and earn almost all the value from their sales, create scarcity in a digital context to maintain prices, easily monitor and report fakes and counterfeits, and collect new royalties each time an NFT is resold, even if it's sold on a secondary market. That doesn't mean any NFT created in the metaverse will necessarily have value; value creation in the metaverse will be determined by supply and demand dynamics for the work on offer. Intrinsic limits on time and creativity will necessarily create rarity, and there will be more and less successful creators. The metaverse doesn't guarantee abundance for all, but it does promise a fruitful ground for value creation, free from intermediaries and gatekeepers, for those with the best skills who achieve product-market fit. Its beneficiaries will include graphic and 3D designers, game designers and gaming companies, natively digital fashion labels and established ones that learn to build in 3D, and an entirely new generation of experts, from virtual real estate developers to e-commerce pioneers who redesign shopping experiences in a 3D context.

To dig more deeply into this last example, the metaverse has the potential to become the next Rodeo Drive or Madison Avenue— indeed, it's already starting to happen. In-person shopping has been on the decline since Web2 ushered massive numbers of users onto the internet. The pandemic rapidly accelerated that decline, as physical shopping became not only inconvenient but actively hazardous. Metaverse pioneers will design 3D shopping experiences that are preferable to users, allowing them to try on clothes, visualize furniture in their homes, and access metaverse-native experiences. User demand will be served by a mix of new, Web3-native vendors alongside prominent traditional and Web2 retailers that are able to adapt their brands into the metaverse. In subsequent chapters focused on marketing in the metaverse, we will discuss how existing brands can distill the DNA that made them successful in their original environments and transcribe it into a Web3 context. Just like the wave of digital transformation in the early 2000s, Web3 transformation along with entry into the metaverse is the inevitable next step for many of their businesses.

The Web3 metaverse is already being settled by a heterogeneous blend of major companies and newcomers. The most popular

Web3-enabled metaverse worlds today are Decentraland and Sandbox; Sandbox was produced by Pixowl, a gaming studio whose parent is the gaming giant Animoca Brands, while Decentraland was developed by an independent team of Web3 enthusiasts. In both worlds, users can buy parcels of land represented by NFTs. Land in these worlds is scarce like physical land. Owning land allows users to construct buildings and invite other users into them. Like countries, these worlds have their own native ERC-20 tokens, SAND for Sandbox and MANA for Decentraland, that enable users to mint and sell NFTs and pay for products and services on the platform. The NFTs are portable off-world and can be bought and sold on open Web3 marketplaces. They can also be moved between worlds, as long as the worlds operate by a common set of standards.

Similar to how the ERC-20 and ERC-721 token standards or the Moloch DAO smart contracts catalyzed massive growth by setting standards for development, open metaverse standards are key to unlocking adoption. If a developer builds a new Web3 metaverse world that is interoperable with other worlds and adheres to common standards for designing assets, users can feel comfortable spending money in that world. Metaverse buyers often have the same mentality as other NFT buyers: that they are investing value in assets that could be resold, rather than consuming value. They also avoid being locked into a single platform. Even if buyers of Web3 metaverse assets decide to stop playing the game or participating in the world where they bought the assets, they can use them in another, interoperable world. In traditional and Web2 games, valuable assets earned or purchased on the platform are typically locked in that platform. Mostly, buyers of Web2 in-game assets are consuming value when they make a purchase rather than investing in assets that could be resold. In some games, assets can be resold for fiat currency off the platform, but the assets can only be used in that game, instead of potentially infinite games and worlds. Large secondary markets do exist for assets native to Web2 games, but these are usually mediated off platform by informal mashups of PayPal and escrow services, and they do not benefit gaming companies, which often actively try to block them.

Web3 metaverse worlds like Decentraland and Sandbox are better understood as towns than as individual games. Similar to a town that attracts more development and an array of thriving citizens, and as a result can offer better services and public goods, these worlds benefit from the network effects of attracting as many users, games, assets, and experiences as possible. Their economic models are designed to align with the interests of creators and incentivize them to add as much quality and variety as possible to enhance their worlds and draw in more users. They are incentivized to keep the barrier to entry as low as possible for creators to build and access the buyers that exist on their platforms. Web3 tokenomics enables them to capture the lion's share of their potential value in a different way: the projects themselves hold MANA or SAND tokens, which are designed to increase in value as more people use them inside their worlds. With more users comes more demand for the tokens, increasing the token prices through supply and demand dynamics. The higher the token price, the greater the total value of the projects' treasuries. They can sell tokens from their treasuries to pay their teams and fund further growth.

Most Web3 projects are transparent about the percentage of total tokens they plan to retain in their treasuries. When their treasuries sell tokens, the transactions are publicly visible on-chain. Projects aim to stimulate enough demand pressure for their tokens to offset the sell pressure that selling from their treasuries exerts on prices in order to avoid tanking prices and upsetting their communities. They often try to sell at strategic moments or in small increments to avoid overly affecting prices, but sometimes out of necessity, they are forced to sell at a suboptimal time. Web3 users pay attention to how projects behave with their treasuries. As with DeFi protocols, a "rugpull" in which founders suddenly sell all their tokens or remove them from the treasury to personal wallets can permanently destroy trust in a project and its team, and therefore their reputations in the industry. Established metaverse worlds, just like the battle-tested larger DeFi protocols, have well-known histories of behaving responsibly with their token treasuries from the perspective of their community members, who care deeply that the project teams avoid crashing the price of the tokens they, too, hold.

The Web3 business model essentially collapses the categories of company, investor, and user into a single economically aligned category called *community*. This is a sharp contrast from Web2, where these three groups are separate, often with adversarial economic interests. The investors in a Web2 company would prefer that the company extract as much value as possible from users, while giving them as little as possible in return. Those who work at the company, especially those that are not significant shareholders, would like to take home all the value they can, instead of distributing it to the investors. The biggest losers in Web2 are the users or customers, who want the highest-value product possible, but from companies incentivized to invest the least amount possible in creating it. The genius of Web3 is that it lays the foundation for business models where the relationship among these three groups is non-adversarial, theoretically leading to better outcomes for all categories of participants.

With respect to creators, companies like Meta are approaching the metaverse with classic Web2 business models. Meta's proposal to charge creators a 47.5% fee to build on the platform is similar to Spotify paying creators a tiny percentage of revenue from music streams, or YouTube paying influencers only a small percentage of the advertising revenue they generate for Alphabet. These companies tax their creators for the privilege of accessing the demand on their platforms. Their economic incentive is to aggregate as much demand as possible on their platforms, which gives them the leverage with creators to charge the highest fees they can and pay them as little as possible.

Conversely, teams building Web3 projects are incentivized to offer the best possible economics to creators to draw more of them onto their platforms, since more and better content on their platforms spins the flywheel of growth by in turn drawing in more users. As has been described, with well-designed tokenomics, more users means higher token prices and more valuable project treasuries, bringing project team, investors, and creators into alignment.

Both Web2 and Web3 economic models work and can be enormously lucrative. The salient difference is economic alignment. This is one of the reasons Web3 enthusiasts have conviction that Web3-enabled metaverse worlds will predominate over Web2 ones. Eventually, we believe, as long as platforms offer good UX, creators and users will do what's in their best economic interest. It's not because Sandbox, Decentraland, or others offer the richest gaming experiences. They notoriously do not, or at least not yet; they've only been around for a few years and are still early in their development. And it's not because they have the most users on their platforms to exert demand pressure on metaverse-native assets. Among Facebook,

Instagram, and WhatsApp, plus Meta, with its prominent brand, easily has access to more users. But just as user growth declined over time on Facebook as users abandoned it for other, more desirable platforms, users very well might follow their economic incentives and migrate en masse from Web2 to Web3. Thanks to better economic incentives for creators, self-sovereign root ownership of in-world assets on the blockchain, and the portability of those assets between worlds, over time, Web3 has a fighting chance to become the economic foundation for the metaverse.

Regardless of how Meta's investments in the metaverse perform, it's likely that many of our interactions and activities will eventually end up in a metaverse context, ideally—from the perspectives of users and creators in particular—built on Web3 economic foundations. As a result, subsequent chapters in this book will recommend that a wide variety of Web3 projects—especially in art, entertainment, gaming, and retail—develop road maps that lead toward a robust presence in Web3-enabled metaverse worlds, in order to get a head start.

PART

2

Web3 Marketing in Theory and Practice

5

Inventing Web3 Marketing

THE BUILDING AT 49 Bogart Street: steps away from the Morgan Avenue stop on the L train in Bushwick, Brooklyn, covered in graffiti and stickers like most of the neighborhood's other low, converted warehouses. When I joined in 2016, this was ConsenSys headquarters, a single room in a mostly residential building with no security desk or formal process for entry. The building's unassuming façade—seemingly worlds apart from the gleaming office towers across the river in Manhattan—would prove irresistible to nearly everyone profiling the company, even appearing as the lead image in a Bloomberg profile on the "crypto world" decamping to Brooklyn.[1]

It was largely here that, as CMO of ConsenSys from 2016 to 2019, I had a front-row seat to witness the emergence of Web3 and an important responsibility to tell its story. Marketing wasn't a common skill set among the core of builders working on early Ethereum. When I first entered 49 Bogart, the environment was far more academic than commercial: most of my colleagues were computer scientists, engineers, developers, and a few key businesspeople. My new colleagues were some of the most brilliant people I'd ever met. Some had technical degrees from top universities; others were self-taught coders and hackers.

My path to joining them wasn't an obvious one. After four years of college, during which I was relatively oblivious to the growing

69

crypto movement, I had gone to work for Arianna Huffington at the *Huffington Post*. Huffington and her cofounders had pioneered a new, digitally native approach to producing and monetizing online content through advertising revenue. Instead of investing heavily in expensive investigative journalism reporters (though it would invest in this later), *HuffPost's* in-house writers would mostly summarize stories broken by other news outlets. They would hyperlink to the original sources and give them credit, but ultimately, if *HuffPost* created a more social media–friendly headline for the story, and attached a more visually stimulating image to it that would appear on social media feeds, then it would draw as many—if not more—eyeballs from Facebook and Twitter to the story on its web domain than the original story, which it could monetize against advertising sales. With this model, called *aggregation*, they achieved higher return for lower up-front investment.

At the time there were few web-native media outlets. Print-based publications were just figuring out their digital strategies as they realized they could get wider distribution by publishing online and socializing content through Facebook and Twitter. The Drudge Report was probably one of the first web-native media outlets, and likely the original inventor of the aggregation model, but it was *HuffPost*, coming only slightly later, that was able to achieve this model at scale.

My first employer also innovated in the field of user-generated content with the *HuffPost* blogging platform. There was an air of exclusivity to being accepted to blog on the *Huffington Post*, and yet, bloggers were not paid for their contributions, even if they achieved massive distribution. Bloggers willingly signed up for this deal, believing the platform would remunerate them with a different type of value: exposure. Appearing on *HuffPost*, the bloggers believed, would help them build personal audiences among whom they could promote a business, advance a social cause, or achieve otherwise beneficial notoriety. Although the vast majority of blog posts were read by few, a small minority of blog posts would earn the most traffic of all the articles on the *HuffPost* website. With many bloggers on the platform and therefore many at-bats, a few blogs would produce extraordinary wins in terms of traffic, and the blog was nearly 100% monetizable upside with almost no up-front costs.

This was perhaps the site's most efficient moneymaker. In 2012, when AOL bought the *Huffington Post* for $315 million in one of the first and still one of the largest digital media exits, unpaid bloggers brought a lawsuit against AOL demanding their share, arguing they had contributed a great deal of the value. The courts summarily dismissed their suit. This was the confirmation digital media outlets and social media networks needed: that it was legally permissible to collect content and data from individuals and monetize it without distributing any of that value flow back to the original creators—one of the core pillars of Web2 business models. Though this approach was deemed legal, Web3 would eventually revolt against it, not in the courts but by designing alternative incentive systems that would better compensate creators.

As a special projects editor working for Huffington, I witnessed the machine for producing online journalism in a digital environment and saw the inner workings of both editorial and the blog. Though some in Web3 are justifiably upset by the shortcomings of Web2 businesses, I'm less interested in criticizing them and more interested in capitalizing on their weakness to outcompete the old models with more creator-friendly alternative solutions. Far from moralizing about uncompensated bloggers, I was filled with admiration for the founder, who was and is genuinely a pioneer and an innovator. It was by watching and learning from her that I first realized that I, too, could perhaps someday start a company.

After a couple of years, I identified an opportunity to further innovate on the media business model by offering creators the chance to be compensated for their work, and I cofounded a news platform called Slant. We invited college journalists, and journalism graduate students in particular, to blog on our platform and would compensate them with 70% of revenue from the advertising we served across their content, taking home 30% for the company. Far from an agnostic platform such as Medium, where writers can publish pieces exactly as they write them, we worked closely with our writers to help them edit and package their content for maximum distribution online, teaching and learning the art of content marketing.

At the time, readers and news outlets were fascinated by first-person reports from college campuses as the topic of sexual assault in college and the burgeoning Black Lives Matter movement gained traction in the popular narrative. As a result of this timing and excellent work by a number of emerging writers, a small minority of top-performing articles were able to achieve significant traffic, sometimes earning a contributor thousands of dollars for a single story. However, the vast majority of pieces earned very little attention, and often, to meet our commitment to pay out a writer 70% of advertising revenue on a given piece, we would find ourselves splitting the dollar, or even splitting the cent. At Slant, we depended on third-party payment processors to facilitate the transactions, and because of the fees they charged on each transaction, we would lose money on low-paying content. Although our business model worked in theory, and we were able to achieve the impressive benchmark of over 4 million page views per month shortly after launch, it did not work in practice because of payment processing. Designing an incentive system to draw in creators with better remuneration, we had innovated a somewhat decentralized micropayment-based business model for media even before the start of Web3, but we were stymied by the intermediated Web2 style of sending payments.

Out of concern for the company's viability, I became obsessed with payments. In New York where I lived, there were a number of technology meetups, and as a cofounder of a small but growing platform, I was able to meet and socialize with other founders. In 2015, I met several members of the early Ethereum team, including Joseph Lubin, Sam Cassatt, Andrew Keys, and Christian Lundquist. Despite the code base being public, it was difficult for me to understand the inner workings of the Ethereum Virtual Machine without a computer science degree. However, it was clear the individuals being attracted to Ethereum were some of the smartest people I had ever encountered. Although I was unable with my media and marketing skill set to perform my own independent diligence on the technology they were building, it was easy to diligence the people. I recognized that they were brilliant.

Not only did Ethereum theoretically solve the problem I encountered in my small business, it also unlocked countless business models and modes of value creation that made logical sense but not practical sense, impeded by the limitations of Web2. Wading through a sea of unfamiliar jargon, I was able to extract the idea that they were building what some called an *Internet of Value* that would evolve the *Internet of Communications* that Web1 had built; where Web1 facilitated the global movement and exchange of information at a scale previously unimagined, Web3 had the potential to do the same for value. (It should be noted value is not the same as money, though money is a common way value is expressed.)

At the time we referred to the movement as *crypto*, but occasionally would use the term *Web3* to refer to the larger construct of a web where all kinds of value can be captured and interact frictionlessly, beyond the financial use case of cryptocurrency or blockchain-based tokens such as bitcoin and ether. Crypto today can be considered a subcategory of Web3 focused on the financial use case. As my interest in crypto grew and the perils of media business models reliant on advertising became even clearer, in summer 2016 I decided to exit my startup and join ConsenSys full time as chief marketing officer. There I would become the first Web3 marketer.

Most were unfazed by ConsenSys making its first official marketing hire. They were focused on writing code, contributing to what they believed could become one of the most important undertakings in software history, and likely their careers. The atmosphere was one of barely contained excitement. We knew we were onto something big.

Others expressed concerns about ConsenSys starting to do marketing. The whole point of designing incentive-aligned economic models in Web3 was that they should be self-marketing. Plus the concept of "marketing" had a veneer of inauthenticity, especially from its context in Web2. The word reminded them of annoying ads on Facebook that no one clicks on, Instagram influencers hawking vitamin waters, exaggerated headlines and articles about the Kardashians. To these thoughtful technologists, marketing seemed more like an evil than a good, the art of using simple and deceptive language to manipulate consumers into buying products they don't need. No one likes the feeling of being marketed to; many are happy to pay subscription fees to avoid seeing advertisements. Software developers especially, who were the target users for many of the first Web3 tools such as MetaMask, Infura, and Truffle, particularly disdain ads. They would rather just inspect an open source code base themselves and draw their own conclusions. Web2-style marketing was culturally incongruous with this group of hackers and academics.

At ConsenSys in 2016, our goal was to build the fundamental tools and dapps that would enable more developers to build using Ethereum. Making Ethereum easier to use would unlock the growth of the Ethereum ecosystem, enabling developers to innovate with the

Web3 substrate and build any application they imagined. ConsenSys brought teams together under one roof with shared funding, legal, and back office infrastructure to help them get to market faster and more efficiently. Because these teams were housed in one incentive-aligned company, they could specialize to deliver the most critical applications rather than compete with each other to solve the same problems. It was a clever solution to avoid redundant work.

Early ConsenSys was also fruitful ground for collaboration. There were relatively few decentralized blockchain technology experts in the world, and almost all of them worked at ConsenSys or the Ethereum Foundation. These experts often contributed wherever they were needed and would work on many different projects at once without official roles or titles, motivated by their desire to grow Ethereum.

Thanks to its critical mass of talent, culture of cooperation, and an electric energy where anything felt possible, early ConsenSys was able to build foundational Ethereum standards, tools, and dapps, many of which are widely popular today. These include Truffle, the original development framework for building Ethereum applications; Infura, the Ethereum infrastructure developers use to deploy dapps; MetaMask, the widely used self-sovereign Web3 wallet; and contributing to the ERC-20 token standard.

ConsenSys brought together teams to build dapps on top of these core tools, aimed at a wide variety of use cases and industries. Among these were Ujo, a Web3 music dapp that would pioneer a novel use of NFTs and inspire others; SingularDTV, a platform to decentralize film and movie production; Gnosis, a prediction market platform whose contracts are now widely used in other dapps; and Grid+, a system that promised to make energy use cheaper and more efficient.

Some of these projects would succeed, while others would disband and send their contributors to work elsewhere within the ConsenSys "mesh," the name for the morphing collection of people and projects associated with the company. If the saying is true that 90% of startups fail, ConsenSys certainly achieved a far better than average track record. But the success of any individual project was only part of the point. My job as a marketer was to help these fledgling startups gain whatever traction they could, while also showcasing each startup as a possible use case for Ethereum. This would inspire others to build on Ethereum and drive ecosystem growth, even when an individual startup failed by other metrics like user acquisition or volume. Obviously, we wanted all our projects to succeed, but our larger mandate was Ethereum.

There was no template for marketing in Web3. As the first official marketing hire at ConsenSys, I learned quickly from my colleagues, especially the few who reacted strongly to the idea of ConsenSys starting a marketing department, that the Web2 approach wouldn't work. Luckily, I didn't have a traditional marketing background and wasn't set in my ways. I'd absorbed important lessons from my time at *HuffPost* and Slant about how content was distributed online, and how that engine worked to capture attention, shape perception, and convert audiences to taking the actions that matter to a company. I was no stranger to social media and influencer marketing, and I recognized the importance of gathering people together for physical events. And perhaps most relevant, as a previous startup cofounder, I knew how to think about starting a business and shaping it over time to fit its customers' emerging preferences.

What I didn't know was very much about Web3. It was immediately clear it would be impossible to design a formula for Web3 marketing without understanding the substrate that is Web3. I needed to get to know the product before I could sell it. I read everything I could get my hands on about Ethereum and lured my busy colleagues out to lunch or happy hour where I would mine them for information. Reassured of my intention to market Ethereum thoughtfully, free from hype and clickbait, they were cautiously optimistic that my unusual skill set could be deployed to help Ethereum achieve its vision, and were mostly happy to spend time with me.

But my goal was to commercialize Ethereum and bring Web3 to a retail audience. In my first interview with Joe Lubin, I told the Ethereum cofounder and ConsenSys founder and CEO point-blank that I wanted to make Ethereum a household name like Starbucks and Major League Baseball. This was earth-shaking technology—not flying cars, but the most exciting thing I'd ever encountered in the world of bits. If we presented it the right way, everyone would care about it. The idea of commercializing Ethereum was controversial among the quiet academics and anti-establishment hackers of early ConsenSys, but Joe nonetheless offered me the position. After all, he had founded ConsenSys to grow Ethereum and bring it to more people. Despite the fact that I was only 25 years old, Joe gave me the benefit of the doubt. I was obviously passionate about Ethereum, and it wasn't like the world's best marketers were flocking to an obscure project whose only reputation was for being inscrutably complicated and getting hacked that one time. Hiring me was an experiment, and I think Joe was curious to see what, if anything, I could do. But I would be given no resources: no new hires and no budget until I proved marketing mattered.

I spotted the first opportunity to do so almost immediately thanks to Andrew Keys, an early Ethereum evangelist and master salesman who was leading business development. Warm and gregarious, Andrew invited me out for dumplings in Shanghai, where we were attending the second annual DevCon, the yearly meeting of Ethereum developers and enthusiasts hosted by the Ethereum Foundation. Talking a mile a minute, Andrew effused about a project he and others were working on: launching the Enterprise Ethereum Alliance (EEA). Months previous, Andrew had cold-called Microsoft to tell them about Ethereum. Two executives, Marley Gray and Yorke Rhodes III, had listened carefully. They were excited about its capabilities and curious to explore use cases for their business. Encouraged by the reception from Microsoft, Andrew worked tirelessly to rally the blue-chip companies that wanted to experiment with Ethereum.

Microsoft, Intel, JPMorgan Chase, Santander, BNY Mellon, and Accenture were among the long list of firms that agreed to launch an intercompany working group in partnership with ConsenSys, several other startups, and academic groups. Its goal would be to set standards for enterprise implementations of Ethereum that could be used by all. This was novel, not only because of Ethereum. These were companies that would otherwise compete or at least, disclose minimal information to each other, suddenly agreeing to band together to build a shared resource. Web3 economic models often incentivize collaboration, and here, in the unlikeliest of places—the cutthroat worlds of finance, enterprise software, and management consulting—the spirit was already taking hold.

By the time Peking duck was served, Andrew had me convinced this was a remarkable opportunity for marketing. At that point during summer 2016, when people googled Ethereum, the results would mostly show them its teenage inventor, Vitalik Buterin. While Buterin and the rest of the early Ethereum technology team were impressive to computer scientists, Ethereum was still a hard sell for enterprise business executives who didn't recognize any familiar names on the masthead. Most of the Ethereum presale investors had been private individuals and small funds. There weren't big brands or many famous investors putting their weight behind the project, which made Ethereum seem niche and unreliable.

Blue-chip companies rigorously vet their choices of software and partnerships, and at the time, Ethereum had no third parties validating its merits that would appear trustworthy to decision-makers who, to make matters worse, inherently distrusted crypto. Ever since the Silk Road case, Bitcoin, the best known blockchain network, had been associated with drugs and violence in the internet underworld. But Ethereum wasn't offering an easy way for dealers to sell pot; it was a new web architecture for creating an Internet of Value. If only a group of trustworthy companies endorsed the Ethereum blockchain, businesses of all sorts would recognize its advantages, overcome their skepticism and indifference, and join the movement.

I would continue the conversation with Andrew and other colleagues over the coming months. We recognized this was an important chance to tell the story of Ethereum in the press, which could help reset the narrative that was established with the DAO hack. Serious companies needed reassurance that Ethereum was structurally sound and wouldn't be exploited again. With household name brands involved in the EEA, we could tell the story of what Ethereum is and why people should care and have it featured prominently in every major media outlet. Millions would discover Ethereum for the first time in this new context.

The *New York Times*, *Wall Street Journal*, Bloomberg, Reuters, and thousands of smaller outlets all published the story at the end of February 2017.[2] Andrew and I had pounded the pavement securing coverage from journalists, explaining Ethereum and the EEA, and coordinating press releases and interviews among the member companies, no mean feat with so many corporate communications departments involved. But the plan worked. After that fateful day, anyone who googled Ethereum learned about Buterin, but also Gray and Rhodes from Microsoft, Julio Faura from Santander, and Amber Baldet from JPMorgan. ConsenSys was widely credited with assembling the group and featured in many of the stories, exploding the name recognition for our business, which was still relatively new.

Investors read the financial press, and sure enough, headlines about Ethereum rubbing shoulders with the most prominent firms in technology and finance brought capital pouring into its token, ether. Ironically, the enterprise standards originally imagined for Ethereum didn't even use ether; they called for private instances of the Ethereum blockchain that wouldn't require a crypto token. It was simply the publicity Ethereum was getting in association with top firms that built investor confidence, sending ether surging above $20 quickly after the announcement. By the first Ethereal Summit in May 2017, ether prices would cross $100, one of the fastest price runups in crypto history.

Joe was pleased with his one-woman marketing department. I had ridden in on Andrew's coattails and proved that marketing mattered. What was exciting to Joe wasn't just brand recognition for ConsenSys or investors pouring into ether—of which he owned a significant stake—it was how the story stimulated network effects that would help Ethereum grow. More investors meant more funding for Ethereum projects, more entrepreneurs raising capital to build on the platform, and more developers hired by those projects to contribute to their code bases.

Our metric for success wasn't the ETH price; it was the number of developers building on Ethereum, in languages such as Solidity and LLL. With more developers building on Ethereum came more at bats to build the "killer dapp" that would drive massive adoption. Almost no developers who started working in Web3 in 2017 would credit their career decision to the EEA launch. To the contrary, many would be motivated by a desire to disrupt the banking industry and disintermediate large corporations. But the EEA launch story began spinning a flywheel of growth whose secondary effects would draw countless developers to Ethereum. (The flywheel is a useful image: a heavy wheel on an axis that at first takes a lot of effort to push, but after a while, picks up tremendous speed—the energy from previous pushes now working with the mass of the wheel such that the wheel accelerates, seemingly turning itself. We'll return to the image of the flywheel when discussing certain momentum-generating aspects of Web3 businesses.[3])

I made my first hires onto the ConsenSys marketing team. The first were Matthew Iles and Elise Ransom. Matthew had a traditional Web2 marketing background and had cofounded a marketing agency with his wife that had successfully marketed well-known retail brands like M.Gemi and Web2 companies like General Assembly. Matthew and his wife had designed the General Assembly course on digital marketing. Coming from the media world with my nontraditional marketing background, I wanted to partner with a leading expert, and Matthew had codified digital marketing best practices and taught them to thousands of students. Matthew was flexible and creative. Far from resting on his laurels, he didn't assume that Web2 marketing would work in Web3, but brought structure, process, and a wealth of knowledge.

Elise, too, was a remarkable addition to the team. She had worked in fintech (short for financial technology, refers not only or even principally to Web3 applications but more broadly to technological innovation transforming or disrupting financial services) marketing for a real estate–focused startup. Though only a few years out of college, like me, Elise was obviously passionate about Ethereum. For months she persistently contacted ConsenSys employees looking for a job. I recognized when we met that she was an extraordinary communicator, with a talent for distilling complex ideas from the world of fintech into clear messages. During her interview, Matthew and I asked her one of our favorite questions: explain something complex that you understand really well, but that confuses other people. Elise flawlessly explained the 2008 financial crisis and the role of mortgage-backed securities, charting the course of the collapse that had inspired Satoshi Nakamoto on the whiteboard of our tiny meeting room at 49 Bogart Street.

Together, Matthew, Elise, and I began laying the groundwork for the first-ever Web3 marketing team, a department that by the end of my nearly four years at ConsenSys would include more than 80 people on teams specializing in public relations, content marketing, social media, community marketing, growth marketing, visual design, and product marketing. As a team of non-engineers, we learned voraciously from our technical peers at ConsenSys, always conscious that we needed to understand the constantly evolving Web3 substrate in order to mold it. Our team would choose the best marketing practices to adapt from Web2 and also derive new, natively Web3 marketing practices from first principles. The technology we were bringing to the world shifted economic control away from discretionary decision-making by groups of people and toward transparent governance by algorithm.

And yet, the process of bringing it to market was all about people. We were successful insofar as we brought in the best talent from Web2 and trained ourselves together on Web3. These included Kara Miley and James Beck, who joined soon after to run public relations; Avery Erwin and Everett Muzzy on content marketing; Kanwal Jehan leading community; David Wu and Dean Ramadan in growth marketing; Elaine Zelby, Brett Li, and Camilla McFarland on product marketing, and so many more. After ConsenSys, they would go on to illustrious careers, mostly in Web3. Camilla, Elise, Kara, and Everett would join me at Serotonin, the Web3 marketing agency and product studio Matthew and I cofounded in 2020.

After years training ourselves as the first Web3 marketing team, bringing many of the most successful early dapps and tokens to market, we hung out our own shingle to bring our best practices to the next generation of Web3 projects: Layer 1s, Layer 2s, Web3 utilities, DAOs, DeFi protocols, NFTs, and metaverse worlds. At Serotonin, we have further refined Web3 marketing and continued to evolve alongside the industry, developing a new specialty in Web3 transformation and coining the term *Web2.5* to refer to traditional and Web2 companies gradually adapting their business models to Web3.

As other marketing practitioners enter Web3, we at Serotonin continue to enjoy the advantage of being Web3 natives, deeply familiar with the Web3 substrate and the Web3 community's history and unique character. That being said, we cheerfully welcome new marketers into the space, because unlike Web2, with its finite pool of business to be competed for by rival agencies and practitioners, Web3 is growing at a breakneck pace, and thanks to Web3 economics, the rising tide of growth truly lifts all boats, stimulating network effects to drive more investment, developers, and users. It behooves us to share our best practices with others in the hope they continue to grow our still-fledgling industry, and also to continue to learn from those newcomers, perhaps the readers of this book, who innovate in molding the Web3 substrate and set new standards for all of us.

Marketing in Web3 is simply the practice of connecting potential users with products or services they want, starting by making potential users aware of them. To be successful, marketers must learn how to stimulate demand for a new product, often starting from zero. They must deeply understand their product as well as their target audience.

This information enables them to construct a marketing funnel that takes potential users through the steps of discovery, engagement, use, and retention. In many cases, a Web3 marketer seeks to convert new potential users from discovery into engagement with an invitation to join a community.

"Community" in Web3 usually refers to an incentive-aligned group of people, typically gathered on a communications channel, such as Discord or Telegram. This group often includes the team working on the project, investors who hold tokens or equity in the project, and users of the dapp or protocol. These categories blend fluidly; for example, a user in the community might contribute to the project's open source code base and receive financial rewards for doing so. Members of the full-time team may come to see this contributor as a regular teammate. A token- or equity-holder in the community might start using the product, and a user could decide to invest. Communities typically share sets of values and interests beyond the mission of the project. A vibrant community can look like a digital social club where members work and play together.

For a marketer, communities offer a pool of potential users to continuously reengage and convert to using new products and features. Ultimately, though, as a community grows, the distinction between the role of the hired marketer and the community member with a marketing skill set can vanish. Just as projects invite their developer communities to contribute to code, they can recruit and reward members with marketing skill sets for contributing to community moderation, content creation, visual design, social media management, hosting IRL (in real life) meetups or events, and ambassador or referral programs. It's efficient for Web3 projects to implement high-quality incentive systems to draw out talent from their communities. Correctly designed, these systems lead to more and better ideation and execution with fewer employees and less back office overhead, while remunerating contributors with valuable rewards.

Web3 projects can eventually become self-governing and self-marketing. With a sufficiently strong community, a Web3 project can progress along its path toward decentralization and dispense with formal employment or even disband its legal entity, because these are no longer necessary. Marketers should consider the design for a self-marketing system from their first day working on a Web3 project, but they also need to know that the road to implementing it may be long. Before a community can grasp onto an incentive structure to take over essential functions, projects need to solve the zero-to-one problem of driving discovery for their product and attracting members in the first place. The following sections will cover the steps required to generate initial traction, stewarding that traction into a self-marketing system, and specific strategies shown to work along the way.

6

Know Your Product and Your Audience

THERE IS MUCH about Web3 and related marketing that is novel and revolutionary. "Know your product" and "know your audience" are not: these are practically the first two commandments of marketing. As with most other commandments, however, the hard part of a timeless rule is actually applying it in a particular context. This chapter examines what it means to know—which is to say, to learn about—your product and audience in the evolving context of Web3.

Know Your Product

Sometimes we as marketers have the luxury of choosing a product to market. The most correlated explanatory factor for our team's success bringing Ethereum to market is the genius of Ethereum. We simply couldn't have achieved comparable results if Ethereum were a less mind-blowing technology. No marketing strategy can replace having an excellent product. However, even the best products can die in darkness if no one pays attention to marketing them. Especially early in a project's life cycle, someone needs to stimulate demand and connect it with supply, otherwise the flywheel of adoption never starts spinning.

If a product is excellent—defined as having a target audience that would use it if they knew about it—marketing hastens the process of creating demand and connecting it with supply. It gives the flywheel a good push or a series of them. But marketing in Web3 is far from a silver bullet. With a low-quality product—one that users wouldn't want even if aware of it—there's no marketing strategy that can make the project succeed in the long run. The product is like a bone, the solid core of a project. But a body that's all bone and no muscle just lies on the floor. All muscle and no bone doesn't work either. To walk around the block, the body needs muscles to attach to bones and apply leverage on them. A project's success comes out of the tension between the solid bones of a high-quality product and the marketing muscle that activates it into motion.

At ConsenSys, our Web3 marketing team was deployed to market any project the company built; at Serotonin, we are extremely selective about the marketing projects we take on. Our goal at Serotonin is to help grow the Web3 ecosystem by driving return on marketing investment for our clients. If we believe a Web3 project lacks a solid product vision, we know that no matter how effectively we deploy our capabilities—our marketing muscle—ultimately the client won't succeed or drive Web3 growth, even if we're able to generate temporary buzz. Thanks to tireless efforts to maintain and grow our reputation as not only the first, but the best Web3 marketing team, we have a robust pipeline of potential clients that enables us to be selective. Unlike traditional agencies, we're not guns for hire, and we won't work with projects we don't believe in. We have the luxury of choosing our partners carefully, and we prioritize them based on what we evaluate as their chance of success based on their product and team. A project with an excellent product is primed for marketing to catalyze step-function growth, and we want to deploy our efforts where they will lead to the most value creation. All Web3 marketers would be well advised to choose the products carefully that they market. With a high-quality product, they are more likely to succeed and establish a positive reputation.

At Serotonin we learned these lessons the hard way. Early on, we signed with a Web3 utility company whose product was almost ready to launch. This was a Web3 version of a popular type of product that exists in Web2. Users of this type of product in Web2 overwhelmingly prefer paying by subscription, but the payment model for this Web3 product was pay-as-you-go. We conducted market research on our client's behalf that suggested Web3 users, like Web2 users, would prefer to pay by subscription to avoid the friction of reupping their accounts every time they ran dry. But the company's product team dug in their heels; they had already built most of the product and preferred to launch it without making major changes. Introducing a subscription would have required overhauling much of the product. It was one thing to want to launch quickly, but even months later, when the product was live but failing to gain traction, the product team continued to dig in their heels despite user feedback about the friction of pay-as-you-go. They contended it was the job of the marketing team to go find users who wanted pay-as-you-go, as opposed to the product team's job to iterate on the product based on early user feedback. Over time, we would see this pattern over and over and flag it as a formula for failure.

The marketing and product teams both need to be responsive and flexible. Marketing teams need to seek out users for the product that exists today. They also must gather data from users about their preferences and feed it back to the product team. Then the product team must iterate based on that feedback.[1] Without a proper feedback loop between product and marketing, otherwise competent product and marketing teams are doomed. We still constantly ask ourselves why products succeed and fail, and we identify and test new patterns.

At ConsenSys, I learned that to market Web3, I needed to deeply understand it. That's why the first part of this book on Web3 marketing is dedicated to presenting the key components of Web3. Anyone marketing a Web3 project should make sure to understand the product. Most marketers aren't computer scientists or developers and can't read the actual code of a Web3 project. This shouldn't stop them from trying to understand how it works as deeply as possible, asking founders or engineers to explain in layperson's terms as many times as it takes. Marketers working on a particularly technical product should consider taking an online coding class like CodeAcademy. I did this during my early days at ConsenSys, not because I wanted to build code, but to understand the vocabulary my colleagues were using and translate it to others without software development skills. The worst thing a marketer can do is pretend to understand the technology behind a product; instead, especially at the beginning, they should find a patient colleague and, in the style of our most naturally efficient learners—namely, toddlers—obsessively ask *why* or *what does that mean* until they clarify the fundamental underlying concepts. Here are some good questions marketers can ask of Web3 projects to become familiar with the technology:

- What type of project is it? (Web3 utility, infrastructure, DeFi protocol, NFT, DAO, etc.)
- What is the core value proposition to users?
- How does its technology architecture support this value proposition?
- How does its technology compare to competitors solving the same problem?
- What, if any, UX challenges is the project trying to solve for?
- What rewards are built into the technology for incentivizing behavior?
- Which Layer-1 blockchain does it use? (Bitcoin, Ethereum, Avalanche, Solana, etc.)
- Does it use any kind of Layer-2 scalability solution? (Polygon, StarkNet, Optimism, etc.)
- How does it address decentralization? Is there a decentralization road map?
- How does it address scalability? What kind of user volume could it accommodate?
- How does it address security? What are the vulnerabilities? Has it been audited?

Though we as marketers need to understand everything possible about the product, this isn't necessarily true of the end users, unless those end users are developers. For B2D (business-to-developer) products, developers usually need to understand exactly how a product works, usually through tutorials, thorough documentation on open source code, and inspecting the code base themselves on a platform such as GitHub. For B2B (business-to-business) and B2C (business-to-consumer) products, the end user often doesn't need to understand *how a product works*; they simply need to know *what the product does*. These are not the same thing. Few people who use the internet understand how TCP/IP works, but that doesn't stop them from checking their email. If being able to re-create a light bulb were a prerequisite to flipping on a light switch, most of us would be sitting in the dark. Knowing how a technology works is essential for marketers, but not necessarily users. It's our job as marketers to be a digestive system for information: to break down chunky technical concepts, extract a clear value proposition, and convey the proposition to its target audience in a way that is accessible and compelling. Sometimes, depending on the audience, that means avoiding blockchain language entirely, even if a product uses the technology—for example, replacing a term such as NFT with *digital collectible.* Just because *we* care about understanding the technology doesn't mean our audience will.

Know Your Audience

Understanding the product is only one-half of the equation to market effectively. The key to success at marketing, and many other parts of life, is to know one's audience. This is hardly unique to Web3. Dale Carnegie's *How to Win Friends and Influence People* was published in 1936. A lecturer who studied the fields of personality and salesmanship, Carnegie was the first writer in the canon of modern business literature to identify a principle that now seems obvious: to be convincing to another person, one must first put oneself in their shoes. Imagining what it's like to be another person enables us to understand their desires and motivations, and then frame our argument or sales pitch in terms that matter to them. When we try to convince another person, the key isn't what *we* want; it's the other person and what *they* want.

A large proportion of the business and marketing books I've read are derivatives of Carnegie's classic, so when a new marketer joins Serotonin, I recommend they go straight to the source and read the original. After nearly a decade of Web3 marketing, we find its essential lesson couldn't be more relevant. As marketers, if we approach our task only from the perspective of the project, we might be able to distill its value proposition into a clear message, but that message is unlikely to be compelling if it's not shaped to fit the target audience receiving it. A good analogy comes from chemistry: an atom must have the right configuration of electrons in order to bond with another atom. In marketing we are trying to forge a bond between a product and an audience. The correct way to do this is study the target audience's "electron structure," custom configure a message with just the right number of electrons, and deliver it over the right channel. Then the message and the audience connect, catalyzing a measurable action, such as downloading a product.

When I first joined ConsenSys, some of my colleagues were worried about introducing marketing to the Ethereum ecosystem. They were familiar with Facebook advertising campaigns, and other intrusive types of marketing that users wish to avoid. Understandably, they didn't want to pollute the thoughtful and academic spirit of Ethereum with unwelcome messaging. But convincing developers to experiment with Ethereum through a Facebook ad campaign would have been an extremely poor marketing practice.

On our Web3 marketing team, we started by learning about our target audience, and studied everything we could about the computer scientists, developers, and entrepreneurs being attracted to early Ethereum. We noticed Ethereum meetup groups for enthusiasts were forming in cities around the world. Their members were college-age computer science students, recent graduates working at startups, and hackers working on their own projects from basements or garages. Some were formally trained, and others had taught themselves mathematics, coding, or economics. Some lived in the Web2 technology capital, the San Francisco Bay Area. Others lived in finance hubs such as New York, London, Dubai, and Singapore. Even more fascinating, meetups were springing up in Budapest, Bucharest, São Paulo, Istanbul, Johannesburg, Lagos, and Jakarta.

Our community marketing team reached out to existing meetup groups and organized a network of meetup group leaders to share best practices. We kept an eye out for Ethereum communities cropping up in new cities, and we launched a program to support new meetup group founders. Through small sponsorships, we ensured meetup groups always had the funds to provide beer and pizza. We circulated decks and templates with standard language about Ethereum and consistent visual design that conferred on nascent groups a sense of legitimacy. As new projects were built on Ethereum, we arranged for their creators to go on IRL roadshows to present their ideas and get feedback from the meetup groups. Group leaders could count on regular, high-quality content, and project teams had the chance to get exposure to early product adopters.

Ethereum meetups grew like weeds. By summer 2017, after the EEA launch and the associated price action and publicity for Ethereum, it was easy in San Francisco, London, Tokyo, Shanghai, or New York to attend an Ethereum meetup every night of the week.

For its most devoted members, mostly young computer scientists and developers, meetups took on a semi-religious air. There were jokes and memes about "the church of Vitalik." But jokes aside, Ethereum meetups indeed provided to many the benefits they no longer found in traditional religion: a meeting place for people with shared values to work together toward a common purpose. We had learned who our audience was (computer scientists/engineers), the right channels to reach them (meetup groups), and the right message, which was offering them tools and resources to build *their own* communities, rather than imposing our idea of what they should look like. The culture of IRL meetups and hackathons as fertile ground for innovation persists to this day.

The lesson of learning from one's audience applies to Web3 today just as it did in 2017. At Serotonin, when we think about growing the community, we start by studying the actions its members are already organically taking and seeing how we can support them. Let's take PROOF Collective's Moonbirds NFT series as an example. When a group for women Moonbirds holders called Ladybirds sprung up, the PROOF team helped its leaders arrange their own online and offline events and supported growing their Discord server by amplifying Ladybirds on social media and letting other holders know Ladybirds existed. Another NFT project, Crypto Coven, noticed that its users were tweeting side-by-side pictures of their IRL selves next to their Crypto Coven NFT witches with the captions "Web2 me" and "Web3 me." Many of the witch NFT holders looked like their NFTs in real life. Others, hilariously, couldn't look more different. The project's founders and team started tweeting their "Web2 me" and "Web3 me." Crypto Coven server members on Discord encouraged each other to share. The meme went viral on Crypto Twitter, strengthening the Crypto Coven community by giving its members their own inside joke and success story. Crypto Coven's founders, similar to the PROOF Collective team and the ConsenSys marketing team, started by observing their audience. Each succeeded in growing communities by learning about the patterns and preferences of their nascent communities and strengthening them, rather than imposing top-down strategies. Web3 marketers don't always assume they're the ones with the best ideas. Rather, they turn to their communities as their source of inspiration and data, doubling down on what already exists.

Communities in Web3 form from both intrinsic and extrinsic rewards. Projects shouldn't hesitate to experiment with both, if they first establish their unique target audience is open to both. Curious scientists and engineers are often attracted to Web3 projects that advance Web3 technology. They may be curious to learn how the project works, meet like-minded people, and exchange ideas. At the same time, they might be motivated by a financial incentive to start experimenting with a given dapp or participate in a bug bounty.

Bug bounty programs are excellent tools for projects to incentivize community members to expose vulnerabilities in their code. Hackers collect a financial reward for discovering a flaw in a project's code and revealing it to the team. Fixing a flaw before the code is live spares projects from painful exploits later. Web3 dapps are considered *antifragile*, which means they become stronger the more they are attacked. Bitcoin and Ethereum are widely considered secure because so many people have tried to hack them for so many years with such a large potential for reward that if it were possible, it probably would've already been done. Newer projects still need to battle-test their code with bug bounties as well as formal audits before many users will trust them with value.

Community members need not be developers or engineers in order to contribute to a Web3 project's success. *Meme contests*, for example, are an excellent tool for communities to incentivize content creation. Thousands of fans are just as (if not more) likely than a small, centralized marketing department to surface what is most funny or compelling about a project or team. Then, even better and cheaper than A/B testing, other community members can vote on the best content. The project, its team, and community are then armed to share the winning memes about the project on platforms such as Crypto Twitter, driving project discovery. In Web3, marketers need not do all the work themselves; instead, they should focus on creating the conditions, often through rewards, for others with the right skills and incentives to work alongside them to achieve their goals.

As communities grow, though, there is a limit to what can be accomplished through extrinsic rewards. The Ethereum Foundation still offers financial rewards in the form of grants for projects building essential tooling on Ethereum that helps grow the ecosystem but doesn't necessarily have an economic model that would be rewarding for the creators. It does not, however, have a blanket policy of paying dapp developers to build on Ethereum as opposed to another Layer-1 blockchain. That is because Ethereum is already the Layer-1 blockchain of choice for the vast majority of Web3 developers, and also because this simply isn't the practice of the Ethereum Foundation, which figures that it can attract developers by building the best technology. Other Layer-1 as well as Layer-2 projects and dapps pay almost any developer to build on their platforms to attract more use cases and activity. This works up to a point. If builders only use a certain platform because they're paid to, as opposed to because the platform serves their use case the best, then they stop using it the moment they stop being paid. Extrinsic rewards for developers at the beginning of a new platform's life cycle can be an effective zero-to-one mechanism to drive early use, but if they don't stick with the platform when the cash dries up, it won't drive long-term return on investment.

One of the best ways to evaluate Web3 platforms is to look at what developers use without being paid to do so, or which have tokenomic models that offer sustainable rewards over time. Eventually, if projects spend more to acquire a user than the value the user adds to the platform, they will run out of fuel. Marketers should take care that extrinsic rewards are actually working to grow their projects without simply creating the expectation for more rewards. They also shouldn't dismiss the power of intrinsic rewards. The early Ethereum meetups didn't work because they offered people beer and pizza. What they really offered was community and identity. Ethereum was a novel technology surrounded by wildly intelligent people. When computer scientists and developers encountered it, they wanted to adopt it as part of their identities. They took up the mantle as local Ethereum organizers and spent their nights and weekends planning meetups. Motivated by fascination with the technology, and willing to integrate it into their identity, their intrinsic motivation often yielded economic rewards. Those with prominent roles organizing Ethereum meetups built their networks and became top candidates for roles at Web3 startups or attractive Web3 founders to venture investors. Marketers should understand how their audience responds to different incentive structures, both intrinsic and extrinsic, over the short and long term.

Knowing one's audience in Web3 marketing doesn't only mean becoming familiar with the existing Web3 community. Consider a

Web2.5 transformation project at a traditional or Web2 company, or a Web3-native project targeting users and industries that aren't yet in Web3—both of these projects' target audiences may be completely unfamiliar with Web3.

Take the example of Audius, a Web3 music streaming platform that Serotonin helped to introduce their token in fall 2020. Most Audius listeners didn't know about Web3, and Audius didn't want to introduce crypto jargon into their messaging. Their users were there for the music, and they wanted to hear about artists and tracks, not NFTs and tokenomics. But Audius also wanted to launch a token by retroactive distribution to reward past users, align them with the team, and build a community going forward. We needed a jargon-light way to distribute crypto tokens to music fans; our job was to message *what they did* rather than *how they worked* on a channel where the message could connect with its audience, while offering a friction-minimizing UX for users to claim their tokens.

So we decided to host a concert. It was the depths of the pandemic, and everyone was stuck indoors. Music fans missing their favorite artists were tuning in to livestreams on platforms such as Twitch. Audius secured electronic music producer Deadmau5 to stream a show on the front page of Twitch. Once a user signed into the livestream, they could watch Deadmau5 spin. On the screen was a banner with simple text message instructions for how to claim their AUDIO—that is, their Audius tokens. The show was a massive hit, and the retroactive distribution of AUDIO earned some artists who had streamed to fans on Audius more value in a single transaction than they made all year in gloomy 2020.

Neither the artists nor music fans were served technical content about blockchains. If they wanted to learn more about the token, they could search or click from the main website to a separate, designated page. This clear audience segmentation with a specific user flow for the crypto-curious enabled the users who wanted token information to access it. Most were satisfied with the *what* of Web3 without the *how*—that is, with claiming their AUDIO tokens and continuing to use the product. Non-Web3 audiences don't care that a product uses blockchain technology. The most important message is the value proposition of the product from their perspective, in this case, getting inexpensive and direct access to their favorite music that they can't find elsewhere. For an energy-focused project, that message might be a cheaper way to buy energy. For a gaming project, it might be access to a favorite game. Most people don't care about the technology underlying the software they use; it's a marketer's job to decide when this is true in Web3.

In our Web3 transformation practice, we even have examples of projects where we removed Web3 language entirely. Serotonin's NFT e-commerce software spinout, Mojito, did a project with the NBA's Milwaukee Bucks in which fans had the opportunity to claim an NFT by retroactive distribution based on their past engagement with the Bucks, such as attending games. We called the NFTs *digital memberships* and left out the word *NFTs*. We and the Bucks agreed we didn't want to make fans feel like we were trying to sell them something; instead, we were merely trying to reward and align them into a community. The digital memberships also really *were* digital memberships. Owning one enabled the holder to access an online community, perks such as discounts on game tickets, and direct ways to engage with their favorite players. Similarly, Reddit launched a collection of NFTs and abstracted the blockchain language entirely, simply calling the series *digital collectibles*.

As NFTs become more ubiquitous, eventually there may be no need to use the word *NFTs*, and we may see more descriptions like *digital goods* or *digital memberships*. A store in the early days of the web would be called a *web store*; today it's simply called a store. As Web3 and the metaverse become the waters in which we swim, we will likely stop calling them by these names, and just call them *the internet* (or *reality*).

At this point in the growth of Web3, it's likely that any new Web3 project attracts a mix of Web3 native users and Web2 users. The Web3 natives will come with self-sovereign wallets such as MetaMask and protest if a platform forces them to use a centralized, custodial wallet. They will expect to engage directly with the team behind the project and to earn rewards for being early adopters. However, Web2 users entering Web3 for the first time won't have wallets. They will fall out of the marketing funnel if they aren't motivated enough to convert through each step of a complicated UX. Used to being treated as consumers, they expect an arm's-length relationship with the project team and may hesitate to join a noisy communication channel such as Discord or Telegram, which are the staples of Web3 communities.

The Web3 natives can be exuberant bordering on spammy. Web2 converts can be intimidated and skeptical. Armed with the knowledge that both Web2 and Web3 users will show up at their project's doorstep, Web3 marketers can put themselves in the other person's shoes and imagine the mentality and expectations coming in from Web2 or Web3. This helps us design messaging and onboarding that suits both of their needs, and accommodates those coming from Web2 along a journey of learning about Web3.

Mojito offers an excellent example. In building NFT e-commerce infrastructure that enables vendors such as Sotheby's to sell NFTs on their own, owned websites (instead of on third-party sites such as OpenSea or Nifty Gateway), Mojito recognized the need to accommodate both Web2 and Web3 users. It pioneered a progressively decentralizing wallet that's easy to set up and takes credit card payments. Although the wallet starts out centralized and custodial as a "multi-sig" (a multi-signature wallet with more than one private key), the user can choose to boot Mojito out of the multi-sig and make the wallet self-sovereign whenever they want. As a Mojito user learns more about Web3, how they use the platform can evolve with them. Native Web3 users can also come with their own MetaMask or other Web3 wallet. Software, UX, and marketing innovations are all necessary to meet users where they are and facilitate their Web3 journeys.

One pattern that has recurred over my time at ConsenSys and Serotonin is that token buyers buy tokens, and NFT buyers buy NFTs. There are obviously exceptions to this rule. Dapper's Top Shot NFT project with the NBA was remarkable in that a significant number of its buyers were NBA fans, as opposed to members of the Web3 community.[2] Other NFT drops by prominent brands have attracted disproportionate numbers of fans of those brands over Web3 natives. Currently, however, this is the exception rather than the rule. Marketers working with brands with existing followings should learn about those audiences, especially whether some segments of them are already onboarded into Web3. Existing audiences that are already Web3 native are most likely to convert to buying a Web3 product. Highly engaged fans of the brand who are not Web3 native can be converted to buying a Web3 product, but unless they are highly motivated, there is usually a higher bar to converting them, including an easy UX for Web2 audiences and clear explanatory content.

Projects with and without Web3 natives in their audiences should assume by default that their buyers are most likely to be Web3 natives, and they should take the time to learn about the Web3 community's preferences in order to tailor products and marketing to them.

Which is to say that, today, most people buying tokens and NFTs do so at least in part for their speculative future value, and are savvy about the factors that make a token likely to increase in price. Those factors include team experience, tokenomic design, community size, rate of growth, previous price history, and project road map. An experienced team with an excellent track record is a formula for attracting investment. Tokenomics should demonstrate how demand is stimulated and the mechanism by which an increase in demand increases price. Metrics matter, like Twitter following and Discord community size relative to level of engagement; low engagement can reveal that followers are bots or community members have been purchased. An ambitious yet credible road map that includes catalysts for step-function growth, such as launching a dapp, metaverse land drop, or fungible token, is also a confidence builder. Projects trying to sell tokens or NFTs to communities that haven't traditionally bought them should be extra careful to verify that these communities would actually buy tokens.

When it comes to moving audiences and understanding how they connect with Web3 projects, we can learn from failures as well as successes. Civil, a Web3 journalism startup that was part of the ConsenSys portfolio, innovated a new business model for incentivizing local journalists. It planned to sell tokens to news readers that, among other benefits, enabled them access to the content generated by its newsrooms. With my history at *HuffPost* and Slant, I was fascinated by this project and sat on its board. As a result, I had a front-row seat to the failure of its token launch. Despite attracting more top-of-funnel token buyers than any other project I'd seen at the time, the UX of buying the actual Civil token involved too many steps for non-Web3 users, and they fell out of the funnel (the Web3 marketing funnel is the subject of Part 3 of this book). Web3 marketers shouldn't count on an audience that's never exhibited a behavior in the past, such as buying a token, suddenly taking that action on launch day. Wherever an audience is being asked to do something new, marketers should test and optimize the funnel to make sure it's actually converting. What happens on launch day should be pretested and never a mystery.

Conversely, I've also witnessed happy stories of projects that assumed they'd attract mostly Web2 users and ended up overrun with Web3 natives. At Serotonin, we worked on the Web3 transformation that earned Sotheby's—founded in 1744, one of the world's oldest and best-known auction houses—inclusion on the *Time* 100 Most Influential Companies of 2022.[3] Sotheby's saw the promise of NFTs early on and decided to make Web3 a core part of their business. Powered by Mojito and with support from Serotonin marketing, Sotheby's launched its own NFT e-commerce site branded as Sotheby's Metaverse. They were cautiously optimistic about converting their existing clients, traditional art buyers, to buying NFTs on Sotheby's Metaverse. But surprisingly, the opposite happened: Web3 buyers flooded into Sotheby's Metaverse, many of them engaging with Sotheby's for the first time. Web3 natives would soon start bidding on traditional art items at non-NFT Sotheby's auctions, such as when ConstitutionDAO bid on a copy of the US Constitution. Very quickly, Sotheby's decided to start accepting auction bids in ether, and even built a replica of its New York auction house in the Web3-enabled metaverse world Decentraland, where virtual attendees anywhere the world can watch its live auctions and bid. With our support, Sotheby's benefited immensely from learning about its Web3 audience.

The best way to learn about your target audience is to observe them and ask questions. Here are some key questions to establish who your audience is and how to appeal to them:

- Who has the problem that your product solves?
- Do they already know they have the problem?
- How are they solving the problem today?
- What are the flaws or shortcomings of the current solution?
- What value proposition of your product resonates with them?
- On what platforms is this audience currently active?
- Is there a particular type of content or signal they pay attention to?
- What other products do they like, and what do these have in common?
- What other communities or groups is this audience part of?
- Are they allergic to any particular channel or messaging style?
- How do they respond to intrinsic and extrinsic incentives?
- In what ways are they able and willing to help with a project?

- What kinds of activities is our community already engaging in?
- What do members want from this community? From this product?
- Why do they leave, and why do they come back for more?

The most powerful marketing strategies in Web3 come from deeply understanding one's product and audience. This is a necessary step toward figuring out how and where to connect the two, with the right messaging on the right channels. But first, projects need to figure out who will do this work, and that means building their marketing team.

7

Build a Marketing Team

SEROTONIN CLIENTS ARE often early-stage Web3 projects, where the only employees are the founders along with a few engineers and product specialists. When we first engage with them, one of their first questions is usually about how to build the ideal marketing function. For an early-stage project that hasn't yet achieved product-market fit, it makes sense to stand up a light and agile marketing department to conduct the tests and experiments to probe for market fit. Teams shouldn't plan to invest heavily in marketing until there's a value-capture moment on the horizon, such as a major partnership, product, or token launch, where excellent marketing would catalyze step-function growth for the project. Even before this return-on-investment (ROI) moment is on the horizon, however, projects should set themselves up for success with the right marketing foundation. That means identifying a target audience, distilling the product offering into a value proposition, and messaging that value proposition clearly to the target audience on the channels where they live.

When building a marketing team, a project should think about the channels most relevant for its current stage based on the size and type of audience it seeks to reach. Most likely, an early-stage, pre-product-market-fit Web3 startup should focus on owned and earned channels at the beginning and branch out into paid channels as it grows and gathers data on ROI. The owned marketing channels for early-stage Web3 projects are typically a website, blog, Discord server, and Twitter account. The blog may live in a CMS on the domain or link out to Mirror.xyz or Medium. Some websites will include a sign-up box for an email newsletter. Events such as inexpensive hackathons and meetups are a live extension of organic social and community channels.

The earned channels for an early-stage Web3 project are media coverage and event speaking gigs. Media comes in many forms, from podcasts to niche media channels to the major national and international media. Virtual and IRL events are common, and founders with interesting projects or well-connected backers can usually swing speaking gigs without paying for sponsorship. Paid channels—such as major event sponsorships, hosting conferences, paid influencer campaigns, media sponsorships, physical advertisements, and paid social media campaigns—should be introduced only once a project can measure their ROI. This is usually impossible without any product or token in market. Over time, projects can measure their customer acquisition cost precisely and optimize paid campaigns so that cost is as low as possible compared to customer lifetime value, the total earned over time per customer.

If a project has a recurring need for a certain type of marketing, let's say creating weekly blog posts, organizing regular hackathons, or engaging every day on community and social channels, then it needs personnel on board to staff those channels. That personnel can be a blend of an internal team with an external agency or agencies. The benefits of growing a team internally include fostering a deeper knowledge of the product. The risk is incurring excessive overhead cost and overstaffing. Early-stage marketing is experimental, and it's hard to know what marketing specialties will be most important. It can also be difficult to find and train excellent Web3 marketers, while select agencies can spin up teams nearly instantly.

From what we've observed at Serotonin, the ideal early-stage marketing department is a mix of at least one internal employee who is a generalist, Swiss-Army-knife marketer, partnered with an external agency. The internal employee serves as a counterparty for the agency, giving feedback on work, providing information about the product, and securing approvals for materials to be released. This marketing-focused employee can also represent the project at IRL and digital events.

As the Web3 project grows and progresses toward product-market fit, it learns which channels work the best and which more-specialized marketing skill sets it will consistently need. At this point it makes sense to increase internal marketing hires, while continuing to use an agency for intermittently necessary specialties, such as media relations and event organizing, as well as extra support scaling the channels that are working the best. Growing projects continue to retain agencies focused on channels such as content, PR, and paid growth to complement and flexibly expand the skill sets of the existing in-house team.

A high-quality agency partner shouldn't feel "external." All contributors should be passionate, incentivized members of the team. Web3 is ushering in a fluid future of work in which it's natural for individuals to contribute to multiple projects that interest them. Some Web3 startups have full-time employees, many of whom also work nights and weekends on their own Web3 projects. Other Web3 projects are structured as DAOs, with no employees at all, and everyone a contributor. The structure of an individual's employment doesn't necessarily correlate with their degree of motivation or the value of their contributions. In Web3, the spirit of trudging home at the end of the workday, or working a 9-to-5 job just for a paycheck, feels like a depressing anachronism. Some of the most exciting talent have left jobs in TradFi and Web2 for Web3 because they believe in the values of the movement and wish to help propel it forward. That spirit infuses the best Web3 projects.

This isn't to say that Web3 startups are without their own problems. Markets that are open around the clock, never closing for nights, weekends, or holidays; pressure from investors; endemic volatility in the crypto market—all this can lead teams into hyperdrive and burnout. DAO structures can be unruly and slow to make decisions. As the founder of a Web3 agency, I'm obviously biased, but I believe agencies like Serotonin can help solve problems like these: relieving pressure on internal teams by adding capacity, and cutting through the noise of a decentralized organization to turn decision paralysis into action. When I was CMO of ConsenSys, I searched tirelessly for the right partner to create the ideal balance of in-house team and agency that one finds in most mature marketing departments. That I could never find an agency that natively understood Web3 was one major reason for starting Serotonin.

When projects choose to hire marketers in-house, they frequently ask us whether they should look to hire marketers who already work in Web3 or train new talent from Web2. The answer isn't simple. When I hired the original ConsenSys marketing team, I didn't have the luxury of choosing between Web2 marketers and those who already had proficiency with Web3, because there weren't any of the latter. Today there are thousands of marketers who work at Web3 projects and companies. In a crypto bull market, the price of Web3 native talent can soar, with venture capital funding flooding the space and projects hiring at breakneck pace. When crypto prices eventually fall, high-quality talent can become available at more affordable rates for startups. In the meantime, projects considering hiring Web2 versus Web3 marketing talent should ask themselves whether they have the capabilities to train someone on Web3. Many firms simply don't have time to teach a new person about Web3.

At Serotonin, we now hire marketers from Web2 because we've developed an internal training and credentialing program to help Web2 marketers adapt their skill sets into Web3. This book includes many of those techniques. That said, the average Web2 marketer won't thrive at Serotonin or in any Web3 marketing role. We only hire marketers from Web2 who are passionate about Web3: who have joined DAOs, Discord communities, and other groups; who spend their personal time, nights and weekends, researching DeFi or NFTs; and who are willing to put in the time and effort to get up to speed. Identifying those individuals is a judgment call, and it's one of the most important ones we make, not just in order to access larger pools of affordable talent, but also because the marketing minds that got us *here* won't get us *there*.

The best Web3 marketing practices are probably yet to be invented, because most marketers are still working in Web2. One important way we help grow Web3 is to open up superhighways for talent to migrate from Web2 to Web3. The same way investors began migrating their financial assets from fiat currency into bitcoin over superhighways such as Coinbase starting in 2012, another type of asset transfer is taking place today. Individuals starting their careers or positioning themselves within companies are learning about Web3 and raising their hands to become "the blockchain person" or "the NFT czar" within their organizations and communities. Though crypto prices rise and fall and investors move in and out of bets, when talented people put in the work to learn about the ecosystem and seek titles branding themselves as Web3, they're often here to stay. It's like the story of the Ethereum meetup organizers; Web3 becomes part of one's identity, which is its own personal type of staking mechanism. One key way we grow our movement is by introducing the next generation of professionals to the substrate that is Web3 and the opportunities it offers. Any Web3 project can join this mission by hiring and training talented professionals from Web2.

8

Set Key Metrics

AT THE BEGINNING of every project or marketing campaign, marketers and their teammates should define success according to clear metrics. These should keep teams focused on achieving outcomes, rather than overthinking inputs. For example, "Publish three blog posts this quarter" isn't a very good marketing goal, because it could be achieved without driving meaningful results for the project if the blog posts don't get distribution and no one reads them. The point of a blog post is to drive traffic to a website or to convert users to download a product or join a community. The blog post isn't published for its own sake. With a goal such as "Publish three blog posts this quarter," teams start to confuse inputs and outputs. Eventually they'll wonder why marketing isn't working.

A better goal is "Bring in 10,000 website visitors from blog posts this quarter." This is what we call a SMART goal: specific (not vague, clearly identifies what we want to accomplish and the steps involved); measurable (includes quantitative/qualitative metrics for determining success); ambitious (not too easy); realistic (not too hard); time-bound (a definite period of time in which to assess/accomplish the goal). There are many different but related versions of what the letters in SMART stand for with respect to goals; these are the ones we use. This type of goal fits into the objectives and key results (OKRs) framework we use at Serotonin to articulate goals and the measurable outcomes that demonstrate achieving them.[1]

Here are a few examples of good marketing OKRs that are SMART goals:

Objective: Make our DeFi protocol the industry leader by the end of Q4.

Key Result 1: Achieve a higher TVL (total value locked) than competitors x and y.

Key Result 2: Achieve a higher total number of transactions than competitors x and y.

Key Result 3: Grow Twitter following by 100% while maintaining the current rate of engagement.

Key Result 4: Grow the Discord community to 10,000 members while maintaining quality, defined as community members interested in participation, not merely rewards.

Objective: Introduce our collection to the NFT-buying community in March.

Key Metric 1: Sell 10,000 NFTs at a 0.5 ETH mint price by the end of the month.

Key Metric 2: Convert at least 2,000 NFT holders into joining the Discord community.

Key Metric 3: Appear in March on two NFT industry podcasts or media outlets.

Key Metric 4: Host a Twitter Spaces discussion of the project attended by 200 listeners.

Objective: Use our fundraising announcement to drive use of our dapp when it launches.

Key Metric 1: Set up a marketing funnel so new potential users who discover the brand from the fundraising announcement can easily and with minimal friction convert into using the dapp. (Dependency on product team: the dapp must be ready on time for the announcement.)

Key Metric 2: Secure a top-tier exclusive media partner to cover the fundraise announcement.

Key Metric 3: Secure at least five follow-on stories and podcasts from the crypto media.

Key Metric 4: Drive at least 10,000 users into the dapp in the month after the announcement.

Objective: Grow our community to take over certain responsibilities from our team by end of Q2.

Key Metric 1: Train and pay ten community members to work as community moderators.

Key Metric 2: Incentivize community members to build at least three dapps using our protocol.

Key Metric 3: Train five ambassadors who can help participants at our IRL hackathon get questions answered about how to use our code, so that we only need to send two team members.

Key Metric 4: Launch a bug bounty program that incentivizes the community to search for bugs.

The most common marketing goals for Web3 projects are growing their audiences or communities and converting them into using or buying the product. An audience or community can be retargeted every time a project has a new product or feature, and in time become the project's most passionate evangelists, contributors, and referrers of new opportunities. Good marketing OKRs should cover success at every step of the marketing funnel. For example, there should be OKRs about using top-of-funnel activities such as media and content to drive discovery, and also OKRs about mid-funnel activities, such as hosting a hackathon for community members to start building using the project's technology. In general, before finalizing OKRs, marketers and their teammates should ask themselves, "If we hit every one of these targets this quarter, will we be successful?" OKRs should be a road map for success. We can build them by mapping backwards from the successful outcomes we want and breaking down the steps to get there.

Web3 marketing metrics can be both quantitative and qualitative, as long as quality can be clearly defined. For example, many Web3 projects get obsessed with growing their Twitter followings and Discord servers at all cost. These projects will often buy followers or community members, or launch giveaway programs that airdrop free tokens or NFTs to anyone who joins. This is often a trap for marketers, because it brings people into communities who don't actually care about the project, and who will often leave once they collect their short-term extrinsic reward, leaving the project no better off for the investment. Vanity metrics such as a high Twitter following and large Discord community will impress some venture investors, but smart ones know to compare the size of following with the amount and quality of engagement. If a channel has 100,000 members or followers, but each post only gets one or two engagements, a smart investor or potential community member can easily tell that the followers or members aren't real. If a project's channels are full of spam, that also signals a poor match between the project's actual offering and the early community it has attracted. Projects should be careful about whom they bring into their early communities, because unless they want to start over and launch new channels from scratch with zero members on them, projects are stuck with their communities for life.

Marketers shouldn't be like the proverbial paperclip maximizer AI, so focused on surface-level growth metrics that they lose sight of the actual point of those metrics: to grow real community around a project.[2] The quality of the content a project publishes and the people it brings together will have an enormous impact on its long-term success. Web3 marketers shouldn't sacrifice quality for quantity. They should do the real work to identify audiences who are interested in their products (or iterate on their products until they fit audience demand) and develop the correct messaging to attract those audiences into communities through testing and iteration. Thoughtfully designed rewards programs can work, but buying followers with low project fit permanently ruins channels.

Goals should be ambitious but not impossible. Usually at the end of a quarter, Serotonin is happy if we hit about 85% completion of our marketing goals. Obviously, 100% completion is better, but if too many of our goals are 100% complete, we start to suspect we made the goals too easy. Many of us are fixated on getting an A+ or 100%, but it's much more important to aim high. That being said, setting goals too high is demotivating to marketing teams. It's rare that a new project with 10 Twitter followers will have 100,000 Twitter followers in a few months. Don't agree to unrealistic marketing goals. Instead, do the work to research what's possible by tracking previous baseline metrics and growth rates and comparing to similarly sized projects.

Once a marketing team is in place that knows its product and audience, and has defined its marketing goals, it's time to design a Web3 marketing funnel.

PART

3

The Web3 Marketing Funnel

9

Introducing the Web3 Marketing Funnel

THE ACTIVITIES THAT constitute Web3 marketing are fundamentally different from Web2. Imagine a marketing campaign in Web2 for a piece of software or a consumer packaged good. The campaign probably consists of paid ads using the Business Suite on Facebook, Instagram, Google Ads, and maybe Twitter. The marketing team uploads the creative (that is, the visual elements of the ad), writes the copy, funds the campaign with a credit card, and selects audiences they want to reach. They probably start small, testing several combinations of creative and copy to see which converts users best off the social media platform to the product's website. They A/B test to optimize the website to send visitors to the cash register, then scale up spend (marketerese for "the amount of money invested") with the best-performing creative on the highest ROI channels. With a Facebook or Google pixel—a tiny image, usually as small as one pixel tall by one pixel wide, hence the name—on a website that can be used to track user activity on the website, they can easily gather data on their target audience and retarget visitors. Using Google Analytics and other platforms such as Mixpanel, they constantly monitor each step of the process, addressing any blockages.

When I started hiring a marketing team at ConsenSys, most of the traditional marketers I met had precisely this skill set. They were obsessed with testing and optimizing paid campaigns. They spoke in numbers and abbreviations and spent their professional lives staring at Facebook ad managers. In a world where marketers have uncertain value to their organizations—since attributing credit for making a sale can be difficult—Web2 marketers tend to present their craft in scientific-sounding terms to appear credible. In my opinion, this misses the point. The numbers and analytics in marketing are descriptions of reality, not the reality itself. Compelling stories and images drive users to take action, then the metrics retroactively describe how well the story worked. Web2 marketers have lost the plot and forgotten about the primacy of storytelling in a sea of figures.

Unsurprisingly, the traditional marketers I met when hiring for ConsenSys were confused about how to bring a decentralized technology ecosystem to market. New Ethereum users weren't going to be reached with a paid Facebook campaign. Determining the customer lifetime value of a developer learning Solidity and comparing it to the cost of acquiring them as a "customer" in order to budget spend on a campaign to achieve ROI for Ethereum (whatever that would even mean) was a nonsensical approach. To add insult to injury, Facebook and Google Ads, the major Web2 marketing platforms at the time, had banned using crypto terminology in paid advertising. If their algorithms detected Web3 language, so much as the word *token* or *blockchain*, in creative or copy, the ad was automatically flagged and removed. In 2017, Web2 platforms from Facebook to MailChimp were so terrified of being accused of marketing unregistered securities that they refused to take crypto companies' marketing dollars.

I was delighted. We wouldn't pay the Web2 pipers to market Web3. Without these data platforms, Web3 marketers were free to return to our creative roots in storytelling and growth hacking. Web2 marketing is a scheme by giant data platforms to keep vendors reliant, so they have to keep paying every day to continue reaching their own audiences. Instead of the Web2 marketing approach, where companies must continue running paid campaigns on third-party platforms forever to keep growing revenue, and revenue is merely an arbitrage between the cost of acquiring a customer and their lifetime value to the company, we focused on using the substrate of Web3 to design self-marketing systems where aligned communities of users, builders, and investors are incentivized to grow projects together. Web3 marketing is like staking a newly planted tree. The stakes are temporary; once its roots are established, the tree should stand up on its own. But fully self-marketing Web3 projects don't spontaneously emerge and market themselves. The job of a Web3 marketer is to start the flywheel of growth until a community can take over. To start spinning this flywheel for a new Web3 project, marketers first need to bootstrap demand.

Sometimes demand already exists and needs only to be directed to the platform where it can meet the supply. More often, though, it's contingent on the Web3 marketer to stimulate the demand for a product from zero, then channel it into a structure that transforms it into economic value. This formula for business growth predated the web. The best way to envision it is as a funnel. At the top of the funnel is *discovery*, the first moment where a potential user encounters a product. The moment of discovery is the marketer's opportunity to present the potential user with a memorable, differentiated value proposition that ideally creates its own category (more on this later). If this message is delivered correctly and resonates with its audience, the potential new user reaches the second step of the funnel: *engagement*, where they actively explore using the product. If all goes well, the potential user will convert to actually *using* the product.

After conversion to use, the marketer's task still isn't complete. Products must not only convert users but *retain* them, or else they risk losing network effects or defensibility to competitors. A broken funnel can leak out of the sides and bottom, while a well-designed funnel is watertight and smooth, converting users through each step with minimal friction. The steps of the marketing funnel are strung together by successive catalysts marketers refer to as *calls to action* (*CTAs*). At each CTA (for example, "click to download") the user chooses whether or not to take the next step. CTAs are ideal points to measure whether the funnel is working. At each step, marketers combat the forces of indifference, skepticism, and inertia, which can prevent marketing funnels from forming or break them later on. A marketing funnel can be working overall but have a break or leak at one step where users fall out or get stuck.

The marketer's job is to design the funnel, monitor the CTA catalysts at each step, and if they're not working, decide whether to patch individual leaks or redesign the funnel entirely. Patches are often experiments with new messaging, a different channel, or even a different target audience. Tests should be conducted with as little economic investment as possible, until a channel has been proven to deliver ROI. Then marketers should scale up investment. This is the modern marketing discipline. Now we can dive into designing a Web3 marketing funnel (Figure 9.1).

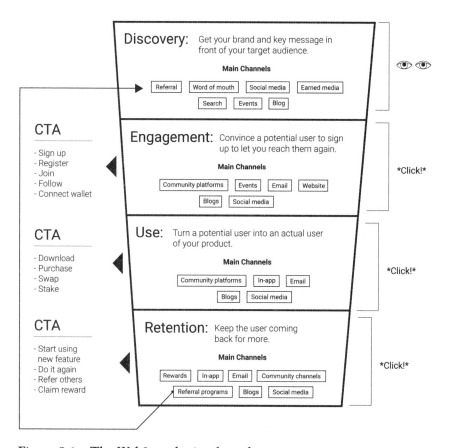

Figure 9.1   The Web3 marketing funnel

Note: Other guides to marketing propose complicated funnels that involve more steps. These steps are probably sufficient for our purpose of understanding marketing in Web3.

10

Discovery

Break Through

Discovery is the top of the marketing funnel, the potential user's first encounter with a product that leads to all subsequent interaction. In this chapter, we'll look at the motives that precede discovery, as well as the messaging and means by which a Web3 product can break through in a crowded marketplace and media environment.

The Origins of Demand

Even before discovery, potential users may already know they need a particular product or service. If a user wants to buy an NFT from a collection on OpenSea, the UX of OpenSea suggests that the user needs to connect their Web3 wallet. The user now knows—if they didn't before—they need a wallet. The next step is choosing a wallet to download. A platform like OpenSea may recommend a short list of top wallets, such as MetaMask, or the user might be left on their own to discover one. They might Google the phrase "most popular Web3 wallets," which yields Google Ads–sponsored results by companies hoping to attract users for their wallets and media sites featuring lists of top wallets. Currently MetaMask tops almost all of these lists, along with Rainbow, Trust Wallet, Coinbase Wallet, and MyEtherWallet. The user might conclude that MetaMask is the most trustworthy because it is the most common result. Then they would proceed to the MetaMask website and follow its prompts to download the Chrome extension on Mac or Windows if they're on desktop, or the app if they're on mobile, and add the extension to their Chrome browser.

This example shows the marketing funnel at work. At the top of the funnel is discovery, or the initial moment they first learn about MetaMask, in this case, through OpenSea's suggestion or Google Search. The next step along the funnel is engagement, when the user visits the MetaMask website. After visiting the MetaMask site, the user takes the next step in the funnel—converts to start using the product—by downloading the extension. Finally, at the bottom of the funnel is retention, when instead of downloading MetaMask and abandoning it, the user returns to MetaMask over and over each time a Web3 wallet is necessary. Wallets are a particularly sticky product, because once a user has funded a Web3 wallet with tokens, it's least frictional for them to continue using the wallet with the assets already in it, as opposed to downloading and funding a new wallet. However, the user may decide to switch wallets if they aren't serving their use case. This can happen if a platform they want to use doesn't accept their type of wallet, or if the wallet lacks a desired functionality, such as displaying NFTs. At the bottom of the marketing funnel, users are still at risk of falling out.

Alternatively to the OpenSea/MetaMask example, potential users might not know yet that they want a product or service. Perhaps the user isn't actively conscious of having the problem that the product solves. For example, imagine a head of finance at a Web3 startup who compensates employees partly in tokens. She may consider it a normal part of her job to manually execute each transaction on-chain herself. A new product such as Franklin, a Web3 native payroll platform that spun out of Serotonin's product studio in 2022, would have to reach her and convince her she has a problem in the first place. This head of finance isn't already Googling "crypto payroll"; a marketer for Franklin needs to find another way to reach her. Perhaps she follows other heads of finance on Twitter, the most popular social media platform in the Web3 community, and notices they're tweeting about a new product that's cutting down the amount of time it takes to do their work by automating payroll. Or possibly she reads a Web3 media outlet such as CoinDesk that publishes an announcement about Franklin raising venture funding. Maybe someone she knows sees the CoinDesk article and sends it to her, thinking, I bet this will help her get work done faster. She could also attend an industry conference and see a presentation from Franklin. Whether the head of finance discovers Franklin through social media, content, the press, or an event, the top-of-funnel message from Franklin should serve to convince her she has a problem ("it takes longer than it should to execute payroll in crypto") and it offers a good solution.

Toothpaste is a famous example from the advertising history of a brand convincing users they have a problem and promising their product as the solution. In the early 1900s, almost no Americans used toothpaste. A friend of the advertising executive Claude C. Hopkins approached him with an idea for a novel paste for cleaning teeth that promoted overall dental health; he called it *Pepsodent*. Hopkins launched a marketing campaign that alerted people to a "film" on their teeth that, he claimed, caused decay and thus less beautiful teeth—a film that could only be cleaned off by brushing with toothpaste. Suddenly, when people ran their tongues over their teeth, they began to notice this film that had always been there, but to which they had never paid attention, much less thought of as a problem. Every time they noticed the film, this would become a trigger for them to practice a new behavior—brushing their teeth—to realize the reward of having beautiful teeth. This cycle would go on until toothbrushing became a habit for most Americans, and Pepsodent a runaway international success. Hopkins stimulated demand for a new product from zero by convincing people who had been perfectly and contently unaware of a film on their teeth that this was a problem, and Pepsodent was the solution.[1]

Knife's-Edge Brand Messaging

Even the most successful Web3 projects at one point had to stimulate demand from zero by convincing potential users first that they have a problem and second that the project's product solves it. This process starts from the moment of discovery, the first activity at the top of the marketing funnel. A marketer's goal in the discovery phase is simply to put their brand and key message in front of their target audience. A brand is a memetic package that can include words and images. The best brands are memorable, differentiated, and create new categories. Key messages are distillations of the primary value proposition of a product into simple language. The best are framed in the terms that matter most to the user, often presented as a solution to the user's problem.

At Serotonin we try to create what we call "knife's-edge" messaging. Although products often have multiple value propositions, we press our clients to narrow down the single most compelling feature or aspect of the product from the perspective of the target audience. They often instinctively resist because they want their products to be known for *all* the amazing things they do, rather than be reduced to a single value proposition. However, capturing the attention of the potential user is difficult, and few will stick around to read a laundry list of features. In discovery, we want to use a single, compelling value proposition to wedge open the door to the user's attention. Once that door is open and the potential user is listening, the project has the opportunity to describe further features. For example, Volvo is widely known as "the safest car," though obviously Volvo has value propositions other than safety when compared to other cars, such as having more horsepower than some, or a better sound system than others. But it's easier for a car buyer to remember a single, hyperbolic message—"Volvo is the safest car"—rather than a long list of value propositions. Marketers refining their products' key messaging should focus on the sharpest, most specific, most compelling single value proposition: the knife's edge.

Differentiated

For Web3 projects, top-of-funnel discovery comes from social media, blog content, earned media coverage, event speaking engagements, and Google Search as well in cases where the potential user knows there is a problem and is motivated to solve it. The potential user first encounters the brand and key message on one or more of these channels. Whether a target audience member is scrolling through their Twitter feed, attending a conference, or walking through Times Square, multiple stimuli are competing for their attention. A new brand and its key message must stand out against its backdrop. Marketers can help their products stand out by designing memetic packages of words and images that are dissimilar to what the audience member paying attention to a certain channel expects.

At Serotonin we often remind ourselves that a zebra in the zoo is quite different from a zebra walking down Fifth Avenue. It's the same zebra, but one is far more captivating than the other, because of our different expectations in context. Marketers can stimulate curiosity and interest in target audience members by putting themselves in their shoes and imagining their expectations in the context of the channel where the marketer is trying to reach them, then making sure their brands and messages contrast against the expectation backdrop.

One example from our practice at Serotonin comes out of our work with Vega Protocol, a Web3 derivatives trading platform. Vega's target audience is Web3 native DeFi users, so they hosted an event at EthCC, a popular yearly Ethereum community conference in Paris. While most DeFi startups at EthCC hosted happy hours or meetups indistinguishable from each other, Vega made its event stand out against the backdrop by inviting chess grandmaster Rustem Dautov to face off against attendees. We and the Vega team figured that the target audience of crypto traders probably enjoyed competitive strategy games like chess. At the conference, Vega was top of mind, with attendees swapping stories about how they fared against the grandmaster. By differentiating its message from its peers, Vega successfully reached its target audience.

Memorable

Marketers should work to create brands that are memorable from the moment they are first discovered. This is partly covered by being differentiated, because we tend to remember what attracts our attention the most through sharp contrast with surroundings. Memory also forms through repetition, as we know from our years spent with flashcards and multiplication tables in elementary school. Users often encounter brands for the first time on multiple top-of-funnel discovery channels (social media, content, earned media, events, and search). If each channel reinforces the same message consistently, users are more likely to remember the brand and message.

Sometimes our clients push back when we insist they use the same key words and phrases to describe their brands in media interviews, their website copy, and speaking at events. But although *they* may get bored of saying the same thing over and over, each repetition likely reaches a small percentage of their overall target audience. Even if our clients are craving variety after repeating themselves 10,000 times, depending on the stage of the project, their average target audience member has probably only heard their message a maximum of two or three times. This is yet another example of thinking from the perspective of the audience, not the project.

Marketing with a holistic strategy across all channels is the best way to manage perception. Each discovery entry point needs to reinforce the message of the others. Multiple sources work together in a patchwork to form perception. If marketers only focus on a single channel such as earned media coverage, a beautifully branded website, or posting the best memes on Crypto Twitter, and they ignore the other channels where their target audience lives, their chosen channel is less effective in shaping perception, because audiences will encounter messages on other channels that dilute or contradict the desired message. Marketers must work wherever their target audiences live and carefully manage each channel.

When we started Serotonin, it became the first Web3-native marketing firm. From the beginning we offered support for Web3 projects on all major marketing channels: brand design, media relations, content marketing, social media, community management, events, and growth marketing. There were other crypto- and blockchain-focused public relations firms at the time, but no one offering holistic marketing across the board. We insisted new clients allow us to manage all of these channels in concert with each other, often collaborating with internal marketing teams. Our theory was that if disparate groups ran these channels without communicating with each other, the voice of a brand would come across as a Frankenstein's monster, a grab bag of inconsistent items unappealingly stitched together. If all channels were managed harmoniously to achieve the OKRs in a shared marketing strategy, projects would be more likely to succeed in hitting their goals.

Our insistence on marketing holistically surprised many of our clients, who would sometimes come to us saying, "Can't you just do our PR?" But my perspective is that there's no such thing as PR anymore, or any other single marketing channel. There is only the sea of perception. Potential users encounter brands on multiple channels, and each experience of the brand works in concert with the others to shape their perception. Marketers are well advised to identify all the channels where a potential user could discover their brand and make sure the messaging on each of these channels is consistent with the others, carries the main brand message, and works effectively to convey users toward the next step in the marketing funnel: engaging for the first time.

It doesn't usually work to "just do PR" for a new Web3 project, because a journalist considering covering the project will probably visit its website and Twitter account. If the visual branding on the website is sloppy, or the Twitter account is a wasteland with few followers, tweets, or engagements, the journalist will be skeptical about the Web3 project and less likely to bestow earned coverage. Plus, even if the journalist covers the project despite its poor website and Twitter, a new potential user discovering the project through the coverage would land on the poor website, question the legitimacy of the project, and probably not convert to engagement.

Creates a New Category

In many cases, brands should also strive to create new categories for themselves. Instead of being one of a type of product that already exists, they should carve out new niches and coin new types of products. If a product is just a type of product that already exists, there is probably a leader in that category that is top of mind with a knife's-edge message already in place. When we search on Google, we tend to only look at the first page of results. Similarly, in a given category, potential users only have the headspace for the top leaders. The safest car? It's Volvo. Nothing else comes to mind. As a first mover in DeFi, MakerDAO was for a long time synonymous with token borrowing and lending. Today, despite thousands of lending platforms launching, only a handful of top projects like Aave and Compound come to mind in the category. The first DEX to achieve network effects, Uniswap, has enjoyed advantages as the category leader. Even today, a few years after DEXes first became popular, only a few other DEX projects come to mind, such as Balancer and Curve. On the DeFi ranking websites such as DefiLlama.com, anyone can see the relative rankings of top DeFi projects by TVL and volume. Visitors will notice that a few platforms account for the lion's share of the totals, and that over time, success by these metrics starts to look like a Pareto distribution ("to those who have, more will be given"). The more established leaders that come to mind within a given category, the more that new products in that category need to differentiate themselves, until it's too hard to break through without starting one's own category.

The burden of proof gets higher for new projects targeting audiences who already use similar products. To overcome the friction of switching products, a potential user must be highly motivated, more motivated than when they started using the original product. Today when a new lending dapp launches, potential users ask, "How is this different from Maker?" Or when a new DEX launches, "How is this different from Uniswap?" When a new Layer-1 blockchain launches, developers ask, "How is this different from Ethereum?" Although many Layer-1 blockchains have launched since Ethereum, many backed by large venture firms eager to match the nearly 6,000× returns (at time of writing, with the price of ether hovering about $1,700) achieved by the motley crew of early adopters from the ether presale, few have reached any substantial scale compared to Ethereum.[2]

As the beneficiary of first-mover advantage, network effects, and having defined a new category—programmable blockchains— Ethereum has for much of its existence processed more transactions than all other networks combined, including Bitcoin. Challengers at Layer 1 need to offer something Ethereum doesn't. The other Layer-1 networks that have been most successful, such as Solana and Avalanche, differentiate in how they solve the security-scalability-decentralization trilemma, optimizing for scalability and lower fees. Other Layer-1 networks have differentiated from Ethereum by specializing in a niche use case or target audience. Casper, for example, can be programmed in the computer language Rust. A developer who wants to build on a blockchain and knows Rust, common among Web2 developers, but not Solidity, Ethereum's primary language, can, according to the company, get started faster with Casper.

If a brand category already has established leaders, new entrants will have a harder time competing. One option is to target a subcategory of audience or use case, but often this narrows the project's prospects.

The other, often better option is to define a new category and become the default leader in that category. When marketers develop their brand messaging, they should always consider product category. How long has this category been around? Are there established leaders? And if there are, how can our product create a new and different category?

Marketers will know they have succeeded in creating a new category through mimesis: if other projects start using the same category name. New categories should be named with the product's knife's-edge differentiated value proposition that matters most to the user in mind. They count as new categories when users perceive them as separate product offerings. You already use Pepsodent toothpaste, but have you ever tried floss? If you're a target user for dental hygiene products, maybe there's not just a film on your teeth, but stuff stuck between them. In Web3, we've helped projects define new categories for their products that match what their target audiences care about most and differentiate from other, similar products.

When we worked with Audius to launch their AUDIO token by retroactive distribution, we wondered how to describe the token to an audience that fundamentally cared about Audius and music, not the technical details of blockchain and crypto. Though the AUDIO token was designed with many properties such as governance voting rights that are exciting to Web3 token-buying audiences, branding AUDIO as yet another "governance token" would be uninteresting to experienced token buyers as just another entrant in an established category and confusing or uninteresting to music fans. We decided with Audius to call AUDIO a *platform token*, signaling to music fans they could align at a high level with the Audius platform they loved, and to Web3 native token buyers that the token offered multiple different benefits from access to governance to whitelisting, packaged neatly together in a single elegant design.

When we worked with Republic, a platform democratizing startup investing, to launch their first crypto token, the Republic Note, we observed that their target audience wasn't familiar with Web3. We and the Republic team designed the Note to be appealing to the audience of retail investors who already were using the Republic investing platform. The investors on the Republic platform didn't know about the benefits they could expect from a token, but they were familiar with the benefits of investing in startups, such as profit-sharing and payouts following liquidity events. We decided to coin a new term for the Republic Note and call it a *profit-sharing token* to clearly articulate the benefit we thought would matter most to the target audience. True to its name, Republic Note holders would see upside whenever a Republic startup exited, so that Note holders shared in profits from Republic. The token launched successfully, and a higher-than-expected proportion of existing Republic users converted to buying the Note. Both the AUDIO and Republic Note launch were in 2020, and over the past two years, many projects have referred to their tokens as *platform tokens* or *profit-sharing tokens*. The new categories we created with our partners at Audius and Republic earned validation through mimesis.

Knife's-edge brand messaging should always strive to be memorable, differentiated, and whenever possible (especially if there is a large established player in place) to create new categories.

Search

Once users know they need a certain category of product, such as the Web3 wallet in our previous example with MetaMask, they usually turn to Google Search to learn their options. This is an ideal type of potential user for marketers, because they are already motivated to discover a product. The marketer's goal is to have their product rank as highly as possible when users are searching for their type of product. They must first learn about the origins of demand that will shape the keywords the potential user types into Google. Returning to the MetaMask example, if a user is coming to Search from OpenSea, and the copy on OpenSea directed them to download a "Web3 wallet," then they are likely to type some variant of the search term "Web3 wallet" like "popular Web3 wallets," "most used Web3 wallets," or "most secure Web3 wallets." Marketers should consider which additional terms would recommend their wallet over others for a user in their target audience. Is their wallet the most secure? Is it the best wallet for displaying NFTs? What is the sharpest intersection point between what the target audience member is searching for and the knife's-edge single value proposition of the product? The answer to this question determines the keywords that should be repeated over and over on the website, on blogs about the product, and wherever possible, in earned media coverage.

Multiple channels work in concert over time and through many repetitions to earn project websites a higher ranking on Google. The more searchers click on links that lead to a product's domain, the more highly the domain will be ranked. This is why it's essential for marketers to maintain a drumbeat of owned blog content over months and years that repeats the key messaging from the website and other channels. The content should also be useful and interesting to readers in order to earn the most clicks from search.

We generally recommend that projects integrate a content management system (CMS) into their domain and publish blogs from their own domain instead of sites such as Medium or Mirror.xyz. Publishing on an external blogging platform grows that platform's domain authority—which makes the domain more likely to appear in search results—and not necessarily the domain authority of the project's website. This can, however, be onerous for small, early-stage product and engineering teams who prefer to concentrate on building products over website content. It's not the end of the world if projects prefer to use Medium or Mirror.xyz. Both of these are standard in Web3. They should keep in mind, however, that publishing on one of these platforms doesn't necessarily refer more traffic to their blog posts. These are publishing platforms, not distribution channels. When publishing on these platforms, it's up to the marketer to earn distribution for content through social media, email, and other owned channels, just like blogging on one's own website domain.

Not every strategy for ranking on Google Search needs to be on the product's own domain. Marketers should proactively reach out to websites and media platforms that publish product listings to make sure their products are included. In our previous example, these would be lists of top wallets. Established media companies already have significant domain authority for their websites. Securing earned media placements helps projects rank on Google Search.

Ideally, a piece of earned media includes the name of the product and its most important keywords in the headline. Journalists won't know which keywords a project wants in its headline, and projects don't usually get to decide on the wording of a headline. The best way to increase the chances a journalist uses the desired keywords is to repeat them in interviews and press materials.

It's also ideal if well-established external domains like media websites and other product websites hyperlink to the product's website. Journalists are often open to hyperlinking the first mention of a product to its website. Before a story is published, the best way to optimize the chances of a journalist hyperlinking is to include hyperlinks in the pitch or press release. Though it's best for relationships with journalists to avoid bothering them with requests for changes after a story is published unless something in the story is truly factually inaccurate, it's generally acceptable to follow up with a journalist who has published a positive story with a polite request to add a hyperlink. They may not do it, but in this case, it doesn't hurt to ask.

When marketers play on our own domains, we have full control over the content; when we work on other domains we don't control, whoever controls that domain decides what they publish there. It's our job as marketers to increase the likelihood that what they publish is useful to us by setting them up for success. Our pitches, press releases, and interviews are opportunities for us to portray our projects in the manner that we wish, reducing friction for them to cover us advantageously.

Marketers should pay attention to what percentage of traffic to their websites is coming from Google Search or other sources. If traffic is coming from sources such as media websites, exchanges, or other product websites, they should look at the specific pages referring visitors to their sites and reach out to optimize any website copy with the ideal product keywords and hyperlinks. If there isn't significant traffic coming from Google Search, it could be because new website visitors aren't the type of motivated potential users that come in through search. Another channel, such as social media, might be performing better for product discovery. Or, the motivated searchers could exist, but they aren't discovering the product due to low search ranking or poor match between what the searchers care about and the current product keywords.

A/B tests on Google Ads can be a good way to test different sets of keywords against the search audience. Marketers should beware, however, of using keywords that poorly represent the actual product, otherwise search may work for discovery, but break at the engagement point of the marketing funnel once website visitors realize that the product doesn't actually do what they want. Even if search isn't a primary discovery channel for a given product today, marketers shouldn't despair; they should still build a robust content calendar of owned content that reinforces key messages. Over time and with optimization, search can grow as a source of traffic to a product website.

Marketers should always be vigilant about how their product is performing on search. In the fast-growing Web3 space, sometimes a competitive project will buy a Google Ads campaign using their competitor's brand as a keyword. Even worse, sometimes scam projects attempting to steal users' private keys take out Google Ads campaigns using the name of a legitimate project. This happens frequently for top NFT collections and claim sites for retroactive distributions. Marketers should immediately report scams to Google, alert their communities about scams on all owned channels, and always include correct links in their own account bios and channels. If competitors are buying Google Ads campaigns for a product's brand name, projects should consider at least a small Google Ads campaign with their own keywords and brand name, especially if it believes its target audience is likely to click on a sponsored link. If not, or if a project is too early to invest in Google Ads, it can still compete through organic search.

Ever since Google lifted its blanket ban on crypto keywords for paid campaigns in September 2018—just six months after putting it in place—Web3 marketers have used Google Ads at a small scale, mostly for A/B testing or to defend products from competitors and scammers.[3] Otherwise, organic search is the focus. While Web2 marketers spend billions on Google Ads, most Web3 companies today do not. This is partly because Web3 companies are still typically early stage and haven't determined customer lifetime value and customer acquisition cost crisply enough to justify the ROI of a large spend on search.

The exception is the large CeFi players such as Coinbase, Binance, and the now-defunct FTX and Celsius. CeFi lending platforms and CEXes behave like Web2 companies on search as well as the other typical paid Web2 marketing channels. It's important to understand these are not actually Web3 companies, though they sell products that have to do with Web3. They usually have no Web3-style strategy to collapse the categories of investor, user, and builder into a single incentive-aligned community—indeed, many are listed on TradFi stock exchanges. The end goal of their marketing strategies is to grow shareholder value not to develop sustainable, self-marketing communities over time. (In our final section on Web3 transformation and Web2.5, we will address non- and not-quite-Web3 companies in depth.) As Web3 companies grow, they are increasing their paid marketing spend, but not necessarily on the Web2 paid marketing channels such as Google Ads. It remains to be seen whether Web3 startups will pay the Web2 marketing pipers. Today, most true Web3 projects don't use paid search.

Social Media

Today, driving discovery through social media means working across centralized Web2 platforms such as Twitter, Instagram, and TikTok. Although many in the Web3 space are striving to build decentralized competitors, Web3 marketers are well advised to meet audiences where they live today, knowing they also need to stay agile to take advantage of new, emerging platforms. This chapter covers the main platforms for Web3 projects to get discovered and grow their audiences, including approaches to take accounts from zero to one organically, leverage influencers, and decide whether or not to invest in paid growth.

Crypto Twitter

Love it or hate it, few can deny the central role of Crypto Twitter in Web3. Though Twitter could not be a more essentially Web2 business, today it is unambiguously the main social media platform for the Web3 movement. Crypto Twitter's relationship with Twitter is ambivalent at best, tense at worst. Crypto Twitter users are often Twitter's biggest critics. Cofounder and former CEO Jack Dorsey expressed his support for Bitcoin over other Layer-1 blockchains, angering much of the Web3 community. But at the same time, Twitter was early to acknowledge Web3 and bring its technology onto its own platform, with the introduction of hexagon-shaped NFT profile pictures that enable users to verify ownership of their NFT in a Web3 wallet. Some Web3 community members have NFT hexagons as their profile pictures, but many others choose not to use this feature. Other social media platforms such as Instagram are quickly following suit in integrating Web3 technology, especially NFTs, into their products.

Despite the tension, Twitter is where Web3 lives, and the most common discovery channel for Web3 projects such as DeFi protocols, dapps, and NFT collections. It's also the best place for newcomers to learn about Web3, arguably better than any Web3-focused media outlet or course. At Serotonin, when we hire new teammates from Web2, we encourage them to spend time on Crypto Twitter as part of their training in Web3. Most Web3 enthusiasts tend to condemn algorithmic social media as the epitome of the Web2 business model. Be that as it may, engaging with Crypto Twitter content and letting the algorithm serve them more helps train our new teammates. Until a Web3 platform obviates the need for Twitter, we give the devil his due.

For the time being, Twitter plays a significant role in shaping the perception of potential investors, users, and contributors. When a Web3 native hears about a project, one of the first things they do is check out its Twitter account. A large Twitter account that was started fairly recently is a big positive sign. Size isn't everything, though, and savvy visitors will also look at the content and level of engagement. Spammy content promising followers rewards explains why a project would have a large following, and to an informed observer is a bad sign. A community that forms based on superficial rewards won't stick around and steward the project through long-term growth.

Ideal community members aren't in it for a quick buck; they are interested in actually using the product and participating in the community. Funny memes; regular updates and news about the product, investments, and partnerships; strong opinions; and links to articles, podcasts, and blogs are all welcome on a project's Twitter, as well as retweets and engagements with other relevant accounts. As long as the brand voice on Twitter is internally consistent, reiterates the key brand messaging, and resonates with its target audience, successful Crypto Twitter content varies enormously.

The denizens of Crypto Twitter value creativity and originality. They don't shy away from conflict, especially if they believe a Web3 founder has rugpulled their community. When Luna founder Do Kwon tweeted to encourage his community to hold onto their tokens after the collapse of his project that had suddenly cost them billions, Crypto Twitter summarily executed him in the town square. They wouldn't let him get away with it. Crypto Twitter can be brutal, but occasionally also produces kindness and sobriety: Web3 founders mediating fights between other founders, giving one another credit, reminding the community to be more positive and less tribal, because the rising tide of Web3 lifts all boats.

Crypto Twitter also has subcultures with their own emotional registers. NFT Twitter, a niche with more artistic content and less technical jargon, is known for its quirky spirit of mutual support. By contrast, Bitcoin maximalist ("maxi") Twitter can be hilarious or rough around the edges and forbidding to outsiders, depending on which Layer-1 blockchain or token one prefers.

Influencers on Crypto Twitter hold enormous power. The largest accounts that draw the highest engagement are often individuals' accounts, as opposed to accounts owned by Web3 projects. When I started working at ConsenSys, I encouraged Joe Lubin to grow his own Twitter following and personal brand, and to expect these channels to grow faster than those of ConsenSys. Most of us are more interested in people than products. We connect better with an individual person telling their story. Company Twitter accounts are obviously channels to promote their projects; individual Twitter accounts are more often a place for people to express themselves, and they tend to be more entertaining. Crypto Twitter holds its influencers to high standards. They are expected to consistently provide their followers with value and entertainment. Accounts that simply "pump their bags," or encourage followers to buy the tokens that they themselves hold, are less exciting than accounts that share funny memes, relevant opinions, and takes on the news of the day.

Many of the largest Crypto Twitter influencer accounts are pseudonymous. Users speculate who might be behind them. Some have suspected that the rapper Snoop Dogg himself secretly runs NFT Twitter influencer account Cozomo de' Medici. The NFT influencer Gmoney is so committed to his pseudonymity that he originally covered his face in photos with his signature CryptoPunk profile picture. It's less important who these users are than what they offer their followers. Unlike Web2 influencers, whose business is overtly to get paid by brands to promote products (often enough, regardless of whether they actually like them), high-quality Web3 influencers are expected to only "signal boost," or promote, the products they care about. When brands try to appeal to the Web3 target audience with Web2 influencers, or Web3 influencers who are obviously only in it for the payday and don't use the product, Web3 natives recoil. Marketers are well advised to know their audience when engaging in paid influencer marketing. An audience of Web3 natives demands authenticity and holds its influencers accountable.

As the Web3 community grows, and as traditional and Web2.5 businesses enter Web3, target audiences will increasingly live on social channels other than Crypto Twitter. We recommend projects spin up accounts on the primary social media channels where their target audiences live. For some of our clients, that means LinkedIn, YouTube, TikTok, or Instagram. LinkedIn tends to reach a more traditionally business-focused audience. For a long time, though, it was our fastest-growing social media channel at ConsenSys, as increasing numbers of Web2 businesspeople began researching Web3. If a project is good at creating video content, the discovery algorithm on YouTube benefits projects with Web3 keywords, because YouTube users, who tend to skew younger, often "fall down the Web3 rabbit hole" on the platform, watching multiple videos one after the other. The TikTok audience skews the youngest. Projects looking to target TikTok users must first get to know the territory. They can't simply post the same videos as on YouTube—far from it, because TikTok audiences expect short, fast-moving videos that must be visually gripping from the first few seconds. Targeting audiences on Instagram is especially appropriate for aesthetically driven projects, such as art or PFP NFT collections, or Web3 metaverse wearables companies.

Marketers considering social media platforms other than Crypto Twitter for discovery should keep in mind that their users are likely less experienced in Web3. Non-native users require more motivation than Web3 natives to get through the same steps of a Web3 marketing funnel and product UX. When non-natives are concerned, projects must offer smoother UX and more step-by-step guidance along the way. All Web3 marketers should spend time thinking about the user journey for a non-native coming from Web2, because almost every Web3 project, no matter how technical, ends up as the first stop along *some* user's Web3 journey. The Web3 community as it exists today is so much smaller than Web2 that if a project isn't accessible to a user coming in with expectations shaped by Web2, it can miss out on an enormous portion of its total addressable market. The next 100 million Web3 users will likely discover our ecosystem on Web2 social media channels. This presents a huge opportunity for Web3 marketers to help new audiences enter Web3.

Seeding a New Account

Many early-stage Web3 companies come to us with zero Twitter followers. Some haven't claimed their handles or even finalized their brand name. At the beginning of a project, there is always the zero-to-one question of how to bootstrap a following from nothing. The answer is that new social media accounts must be seeded by existing social media accounts.

At the start of its go-to-market motion, every Web3 project should ask itself, What large and high-engagement accounts on Twitter (or any other social media platform where the project's target audience lives) have any reason to help promote my project? These can be personal accounts belonging to founders, teammates, and their friends and family; the accounts of investors or backers of the project; or the personal and business accounts of other projects that are partnered or somehow aligned with the given project. *Large* is also relative. Not every project will have connections to accounts with 1 million or even 100,000 followers.

Marketers shouldn't ignore the power of micro-influencers or accounts with even a few hundred followers. They should granularly list all potentially aligned accounts on their chosen platform, and organize those individuals into a channel, such as a Telegram channel, where the team regularly shares tweets from the project and asks for support in amplifying them. In Web3, this chat group is often called a *supporters channel*, and is a great way for an early project to efficiently share updates with its incentive-aligned community, ask for help, and generate shares on social.

A project's team, investors, and partners are probably willing to help bootstrap its social media accounts, but if marketers expect them to do this proactively, they don't understand human nature. Marketers should establish a push relationship (as opposed to a pull relationship) with the busy stakeholders in the community, proactively pinging them—within reason—with requests for social media support and other kinds of help. When projects meet interesting new people who ask how they can get involved, an easy answer is adding them to the supporters chat. Then they can be efficiently leveraged over and over to provide all kinds of support.

Endogenous Influencer Marketing

At Serotonin, we call the strategy of listing all aligned accounts and organizing them into a channel *endogenous influencer marketing*, *endogenous* meaning generated from within an organism, as opposed to *exogenous*, generated from outside sources. Most Web3 projects we encounter are economically aligned or friends with people who run fairly large and engaged social media accounts. They usually haven't leveraged these connections and sometimes are reluctant to do so. It feels spammy to ask valuable friends and partners to retweet all the time.

We don't recommend making these asks in a way that compromises relationships; marketers must consider the perspective of the person receiving the ask. If the highly connected person's company is partnered with the project, they probably have an incentive to share news about the partnership to demonstrate how active and connected their own project is. A partner will often be receptive to a request to retweet or share, adding their own comments about why they chose the project. A project's investors or backers have an incentive to share because they want the project to succeed. And most of us like helping our friends and showing that we have friends solving interesting problems.

Marketers should ensure the content they post on social media is actually the type that their endogenous influencers *want* to share, whether that means funny memes, opinions on specific topics in Web3, or recent product and partnership announcements. If endogenous influencers aren't receptive to this type of request, it's probably because of poor match between the content posted and what they want to share on their channels, or low intrinsic or extrinsic incentive alignment with the project, either of which can be improved.

If a Web3 project really doesn't know anyone with any kind of social media following, one way to bootstrap an endogenous influencer cohort is with an advisorship program. Reach out to notable individuals with engaged followings who have a reason they might be interested in the project, soliciting their feedback and floating the possibility of becoming an advisor. Advisors are usually incentivized with a small amount of equity or tokens, and it's helpful to create upside alignment.

But good advisors should also be motivated by something else, such as the chance to establish themselves as thought leaders, a genuine desire to share wisdom with newer projects, or an interest in helping valuable potential business partners enter the market. Projects with and without endogenous influencers at the ready should consider the value of an advisor program. Every Web3 project should engage in the practice of endogenous influencer marketing to take full advantage of the community they already have—but simply haven't organized—on day one.

The World of Women (WoW) NFT collection, 10,000 individual NFTs featuring images of women of a wide variety of appearances, is a perfect example of endogenous influencer marketing in action. The project connected with celebrity actors Reese Witherspoon and Eva Longoria, who wanted to associate themselves with Web3 while also demonstrating support for women and people of color. WoW was the perfect project. Reese and Eva each publicly bought WoW NFTs, joined the WoW community channels, and tweeted about it. The tweets were positively received on Crypto Twitter. If Witherspoon and Longoria had tweeted paid ads—and paid ads must be marked #ad according to most social media platform guidelines, destroying any chance of being perceived as authentic—promoting a DeFi platform such as Maker or Aave, the Web3 community's immune system would have rejected them. But it felt believable that these actors would engage with Web3 through a collection such as WoW. The two celebrities genuinely joined a Web3 community and became its vocal public supporters; the community and Web3 overall benefited as a result.

Incentivized Growth

At every step of the marketing funnel, Web3 marketers fight against the forces of inertia, indifference, and skepticism. For some target audiences, especially Web3 native ones, paid influencer campaigns produce a lot of skepticism. Web3 projects, they think, should be able to attract aligned and interested community members to promote them without having to pay. Plus, an open source project targeting developers shouldn't have to pay to advertise its benefits, when developers can easily diligence the code for themselves. These groups are particularly sensitive to the feeling of being marketed to, and rightfully so, because they are powerfully capable of verifying the truth for themselves.

Facing this type of audience, marketers should be rigorous about only seeking promotion from endogenous influencers who are authentically users and fans, otherwise they do the project a disservice. For other types of audiences, however, and as the Web3 community grows to include more people with expectations shaped by Web2, paid influencers can be impactful. CeFi platforms like CEXes and borrowing and lending companies unabashedly use paid influencers in the manner of Web2 and TradFi companies. High-quality Web3 projects rarely use overt paid influencer marketing, defined as posts with #ad on them. But as the composition and dynamics of the Web3 audience evolves, they may in the future.

One caveat is that attitudes about paid advocacy aren't universal. Although promotions by paid influencers are distrusted by audience members in the English-speaking Web3 markets, they are widely accepted in the Mandarin-, Korean-, and Japanese-speaking markets. Influencers are usually referred to as *KOLs*, or *key opinion leaders*, and are broadly used in these markets to promote tokens, NFTs, and other Web3 projects. A large portion of audiences in these markets discover projects for the first time through their favorite KOLs, frequently on chat applications such as Telegram, KakaoTalk, or WeChat (though less so recently on heavily surveilled WeChat, ever since China's latest crackdown on crypto). In these markets, audiences tend not to expect comments by KOLs or even journalists to be editorially independent from influence by companies or governments.

In certain markets, it doesn't cause skepticism that a message was paid for by a company, because people assume an extrinsic motivation underlies any message. As a former member of the US media (albeit on the business side as opposed to a reporter), I was shocked when I started marketing Web3 projects internationally and realized how drastically the information ecology varies among countries and cultures. I observed that it's a particularly Western expectation that the media and other information sources would be "objective" or "unbiased." (Whether such sources ever actually accomplish this in the West or anywhere else is another story, but I do still find value in the endless work of trying to discern and be transparent about bias and influence.) From a purely marketing perspective, this is another example of the importance of understanding one's audience. Although paid influencer marketing rankles some audiences, KOLs are perfectly acceptable in others. The key to discovery is just that: figuring out what the audience likes and giving it to them, but with a memorable twist.

Earned Media

When Andrew Keys and I briefed the press about the launch of the Enterprise Ethereum Alliance (EEA)—the spark that would light ether price action in 2017—most of the journalists we spoke to were fintech reporters. Industry outlets such as CoinDesk already existed and were starting to gain traction, mostly covering bitcoin prices, but at most major publications there was no established blockchain or crypto beat. As our marketing team grew and we made our first media relations hires, we focused on starting personal relationships with the journalists covering the nascent space. Many of them were personally fascinated by Bitcoin and now Ethereum. They were considering whether to continue as generalist technology, finance, or fintech reporters, or whether to take the leap and define their careers on blockchain and crypto.

We built a small community of journalists who were interested in understanding more about the technology and offered them briefings on background from experts in ConsenSys. This helped them get their technical questions answered. They were in uncharted territory covering complex technical concepts and trying to simplify them for generalist audiences while being factually correct. With our team of Web3 experts at ConsenSys, we were in a position to help. All we expected in return was more accurate reporting about the ecosystem we were building. Over time—and long lunches at Roberta's, the much-loved pizzeria next to our office in Bushwick—we built trust with journalists. As crypto prices soared in the bull market that started in 2017, thousands of crypto projects began competing for their attention, thirsty for legitimizing media coverage by top-tier outlets. Leaning on our long-term relationships, whenever we had an announcement or product to launch, the ConsenSys media relations team could cut through the noise and earn coverage from these journalists, who could trust (without being able to diligence the code themselves) that we wouldn't send them projects that would turn out to be scams.

During that period of exuberance, however, many fraudulent crypto projects *were* able to con journalists into giving them positive coverage. This led to confusion from investors and the public. The result was that most major outlets restricted who was allowed to cover blockchain or crypto news to a designated expert individual or small team of reporters. They became more skeptical about covering any project built on a blockchain, making the standard higher for projects to get coverage. Sometimes when projects wrote to editors, they simply wrote back saying, "Sorry, we don't cover crypto." Thanks to our relationships, our team could still break through the blockade. But our reputations were at stake, and could be lost with one false move; we took our responsibility seriously. At the same time, we promoted Ethereum, ConsenSys, and its projects to the maximum extent possible in the media, with the sincerity of true believers.

When relationship-building combined with newsworthy results eventually does lead to media coverage, there are a number of best practices we advise our clients to keep in mind. We recommend project leaders arrive at any media interview prepared with the number one key message they are determined to communicate, as well as answers to any questions a journalist is likely to ask—ideally, answers that link back to repeating the key message. They should try to state their number one key message at the start of the interview instead of saving it for later. It's hard to predict when an interview will be over, and the only way to ensure with 100% certainty that a message makes it into the interview is to state it as early and as often as possible.

Interviewees should know that anything they say can be taken out of context, and that it's up to the journalist which quotes they wish to use. If the interviewee says anything wrong, the journalist can print it in the story. The only corrections that are appropriate to ask of a journalist at a top-tier media outlet are to correct factual inaccuracies that are not in quotes by the interviewee. We generally recommend not going "off the record" or speaking "on background," and simply being prepared to answer any tough questions with deferrals. In hot water, it's always possible to say some variation on "My understanding was this interview would be about x, and my view of x is . . ." or "I'm not the right person to answer that question, let me get you in touch after the interview with the correct person."

In the rare moments in which a project or individual wishes to communicate information discreetly to a journalist and avoid being mentioned by name in a story, we recommend communicating those messages through a media relations professional who is experienced with off-the-record and background conversations. Some industry outlets in Web3 have poor track records keeping agreements with staying off the record, on background, or waiting to publish until a news embargo lifts. A professional with experience in Web3 knows which outlets fit this description and how to navigate the space.

Trust Issues

Marketers working on media relations in Web3 still operate in an environment of mistrust. Web3 projects have so much to gain economically from positive coverage in top outlets that many are willing to say anything to earn it, regardless of whether information is true. At Serotonin, our long-term relationships with journalists in the crypto and generalist media continue to serve us well. Our partners at these outlets know we would never intentionally steer them wrong, and we do the same level of diligence as any venture capitalist (in fact, most of our clients are referred directly by the leading venture capital firms in Web3, as an extra layer of validation) to verify that the projects we take on offer unique and powerful solutions and operate with strong business ethics.

Aside from relationships, the best way to break through distrust to earn media coverage on behalf of a project is with names and numbers. If a media pitch to journalists includes impressive numbers and recognizable names, it is far more likely to earn coverage.

The names of prominent venture capitalist backers, noteworthy partners, and team members with strong reputations help media pitches gain traction. The EEA launch is the ultimate example of existing brands legitimizing a new brand and earning it coverage. When we pitched the Web3 developer tools and infrastructure company Alchemy, we made sure to drop the names of its prominent investors, a group that includes several large crypto funds and the rapper and music mogul Jay-Z. Perhaps unsurprisingly, Jay-Z found his way into headlines about Alchemy; having such an unexpected backer helped them stand out from other Web3 companies.[4]

When Serotonin pitched the Republic Note to journalists, we focused on numbers. Journalists just learning about Republic were surprised that it already had 700,000 users, an extraordinary number of users at the time and a powerful validator.[5] In the case of both Republic and Alchemy, we were able to land our clients top-tier media coverage, including features in *Bloomberg*, because we focused pitches on the names and numbers that would make a new project stand out to a reporter.

Numbers ideally should be verifiable. If a DeFi protocol can prove on-chain that it has generated a newsworthy amount of TVL, that it gained a significant number of users, distributed its token to a record number of wallets, or yielded meaningful earnings to a number of people in a developing country—depending on the size and impact of the data, this piques the interest of industry or generalist outlets. Without hard data, claims that a project is positively affecting underserved communities, generating massive volume, or producing extraordinary returns will be greeted with skepticism and fail to earn quality media coverage. The same goes for future-looking statements. Newsworthy events are the ones happening now, not the potential for something to happen in the future. Many Web3 projects have ambitious plans. Plans, however, aren't newsworthy by themselves.

One particular number the media pays attention to is token price. Price is a double-edged sword for Web3 projects. When prices are rising, coverage that focuses only on the price can be shallow and neglect to mention a project's technology breakthroughs or larger mission. At the same time, positive stories can be a boon for top-of-funnel discovery. If the coverage of a project focuses on price on the way up, it will similarly focus on price on the way down.

We recommend marketers avoid pitching the media with stories focused on the price of their fungible tokens or NFTs and also avoid commenting on their own project's price on social and community channels. (Most of the community channels we administrate have explicit rules against price talk, and members who talk about price can be kicked out of the channels. This is a great way to avoid speculators and prioritize building a community that is genuinely interested in a project and its technology.) If the media reaches out and insists on covering price, use their interest as an opportunity to tell a broader story about the merits of the project, always returning to the refrain of the brand's knife's-edge key messaging. Use journalists' obsession with price to smuggle in the true value proposition of the project. This can feel unnatural, but the more frequently an interviewee repeats their key messages, the higher the likelihood that they end up included in a story or segment.

Backing into PR

It's easy for marketers to think about earned media from the perspective of the project. *The project has x and y partners, and it is launching z.* A journalist may or may not be interested in such a pitch. Instead, a more effective way to think about media pitching is to think about the target audience for the pitch: the journalist. At a quality outlet, journalists will only cover stories they deem newsworthy. That means significant enough names and numbers for their caliber of outlet. For example, a Web3 startup raising $5 million may earn coverage in an industry outlet such as CoinDesk, Decrypt, or CoinTelegraph, but it won't earn coverage in *Bloomberg* or the *Wall Street Journal*, unless there's something else unusual or extraordinary about the story. A Web3 startup raising $100 million, however, will usually earn coverage in the mainstream media.

If a project wants to target a particular media outlet, marketers should first figure out what kinds of stories that media outlet would be willing to cover, then work backwards to produce the facts that would enable them to pitch such a story. Let's say a project wishes to be discovered by the target audience that reads the *MIT Technology Review*. This publication covers technological breakthroughs of a certain magnitude. If the project wants to land a story there, as a prerequisite it must work on producing such a breakthrough. If a project wishes to get covered in the *New York Times*, it must consider what it could do of a magnitude that would be newsworthy to a *New York Times* journalist.

It's inadvisable to send pitches to outlets and journalists unlikely to cover the story. Journalists like to see that projects have done their research and send them the types of stories they might actually cover. This helps projects foster positive long-term relationships with journalists. The "spray and pray" approach of sending a pitch or press release to as many journalists as possible, regardless of the match between the pitch and what the journalist has covered historically, doesn't work nearly as well as doing the research and cultivating long-term, mutually beneficial relationships.

The more a marketer speaks with a journalist, the more opportunity they have to learn from the journalist what types of stories they would like to cover, allowing them to further refine and customize their pitches. Furthermore, the project stays top-of-mind for the journalist, who may proactively approach the project in the future with their own story ideas. Marketers should offer their project leads to journalists as experts in their fields. Experts can help journalists by adding commentary to stories, even if the subject of the story is another company in the industry. If project leads are notable experts on hot topics of the moment, marketers should reach out to journalists to proactively offer them up as sources for commentary. Like every part of marketing, success in earned media is about understanding one's target audience. Marketers should take the time to understand what journalists at their target outlets are looking for and can even backfit their projects' strategies and activities to match what would earn them coverage.

The Role of Stunts

Journalists at top outlets will cover a zebra walking down Fifth Avenue, but not a zebra hanging out in the zoo. When we work backwards from what an outlet wants to craft a pitch about a project that they will accept, sometimes we find the outlet's standard is too high to be easily achievable. For example, it's almost impossible for a new Web3 project to earn a feature in the *New York Times*. Early on, most projects don't have the names or numbers to earn coverage from top-tier outlets. One way to circumvent this and skip to the front of the line to get coverage is through stunts.

Examples of stunts in Web3 include the Layer-1 Ethereum competitor EOS—which attracted attention in 2017 and is now for all intents and purposes defunct—buying a giant Times Square billboard outside the yearly Consensus crypto conference hosted by CoinDesk. At the time it was unusual and surprising for a crypto project to take out an actual physical advertisement; crypto ads were still banned on Web2 advertising platforms such as Facebook and Google Ads. The billboard became a topic of discussion, and many conference attendees discovered EOS for the first time as a result.[6]

In 2021, the crypto investor who goes by the pseudonym MetaKovan (real name Vignesh Sundaresan) bought an NFT by the digital artist Beeple from Christie's auction house for a whopping $69 million. By playing on the internet joke about the number 69 and overpaying for an NFT by the standards of the time, MetaKovan won notoriety in Web3 that gave him the opportunity to raise his public profile and attract more funds.[7] Similarly, in 2021, the pseudonymous NFT influencer Gmoney paid $170,000 for a CryptoPunk NFT, also considered a shocking overpayment. This moment launched Gmoney's career as an NFT influencer, and he is now one of the most prominent collectors in the space, advocating for the value of NFTs. Gmoney coined the term the *digital flex* to explain why he paid so much. Since that moment, thousands of NFT investors have performed their own digital flexes.

Gmoney created a new category, validated through mimesis. He earned coverage on CNBC and other outlets; MetaKovan made the *New York Times*. If Gmoney had paid a market rate for a CryptoPunk, or MetaKovan had bought the Beeple NFT for a modest sum, neither would have made the news. Both Gmoney and MetaKovan used costly stunts to get discovered, earn coverage, and achieve their goals.

Stunts require real risks and investments. Projects considering performing stunts must fully think through the stunt, the outcome they expect to result, and how to turn the interest from top-of-funnel discovery into lasting value. Otherwise, a stunt is likely not worth it. The best way to engineer a stunt for discovery is to know one's audience, working backward from what would matter, or what would stand out, to them.

Attention Is Jagged

The best moments to pitch for earned media coverage are when a project has the biggest news of the type that outlets most want to cover. These can be engineered by teams through stunts or activations. Others occur naturally in the life cycle of most projects. Marketers should identify the highest-impact moments to help their projects get discovered and focus on executing these properly.

For an early-stage Web3 project, the first pitch is usually a fundraise announcement. This is the first opportunity for projects to anchor their brand messaging with potential users. If a project has raised substantial funds (the numbers) from notable investors (the names), it's likely to earn some degree of coverage. Marketers should use this opportunity to convey the project's key messaging, the names of notable partners, and any traction numbers so far, keeping in mind that with any new project, they're battling against the forces of indifference, inertia, and skepticism.

Other ideal moments to pitch for media coverage include the launch of a major product or feature, surpassing the leader in a category, achieving an important milestone such as 1 million users, the announcement of a significant partnership, and discovering an unexpected use case. These stories are also possible to pitch, usually to industry outlets: hiring a notable person onto the team, expanding into a new geographic region, or hosting an extraordinary IRL event. It's also possible to pitch human-interest feature stories that focus on a single person's (or group of people's) experience with the project. This requires real research and planning.

Unless projects engineer the opportunity with a stunt, the chance to tell any of these media-friendly stories is rare, so marketers should take maximum advantage wherever they naturally occur. These are the best moments to drive top-of-funnel discovery through earned media, social media, and events. If the moment is well executed with knife's-edge messaging on the appropriate channels that resonates with its target audience, it drives step-function growth in brand recognition. Success looks like increased search volume, website traffic, and social engagements and followers. Afterwards, a project's baseline metrics should stick at a higher level.

Attention is jagged and not every moment is the same. Figure 10.1 is the basic graph of what it looks like.

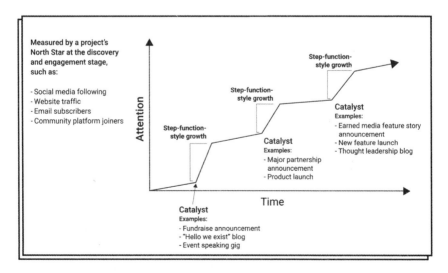

Figure 10.1   The attention a project gets over time is jagged

Chainlink Strategy

Although marketers must thoughtfully optimize the highest-impact moments to drive maximum discovery, they should also consider developing content calendars of continuous promotions. Discovery happens at big-bang moments through major news coverage. It also happens quietly and constantly every day. Matt Coolidge, an early team member at Serotonin, came to us from the prominent Web3 oracles project Chainlink, where he led communications (recall that, like Weather.com in our example of how smart contracts work, an oracle is a source of outside data that is fed into the execution conditions for a smart contract). We were excited to work with Matt for many reasons, one being how much we admired Chainlink's strategy of constantly messaging the market, which had proven extremely effective.

Observing Chainlink from the outside, we noticed that every time a new dapp or protocol started using Chainlink, it was all over Crypto Twitter. Either Chainlink, the new project using Chainlink, or both, would publish a blog announcing the partnership; then both would share and cross-amplify each other's posts on social media. Each blog post reinforced Chainlink's key messaging and grew its domain authority with hyperlinks from other websites. The cross-amplification helped the Chainlink brand reach new audiences, in return, helping the partner's brand grow by reaching the Chainlink audience. As a result, community members and potential investors who were already closely following the project perceived the project to be constantly active. Newcomers who had never heard of the project before discovered it for the first time on a channel where they lived, like the Twitter feed of a dapp they already used. Though Chainlink might have worried it was pummeling the market with constant messaging and annoying followers with too much content, each tweet only reached a small segment of its target audience via algorithm, so that from the potential user's perspective, these messages weren't served up too often. It was ideal for everyone.

We introduced this approach to Serotonin clients and called it *the Chainlink strategy*. Marketers can implement the Chainlink strategy for their projects by standardizing a playbook for announcing new partnerships and executing it each time the project collaborates with a new partner. Over time, this becomes a well-oiled machine for growing brand recognition. Social media followings and engagements increase, along with domain authority for search rankings. Marketers can overcome the hurdle of skepticism and foster trust in their target users by demonstrating how many other, similar projects are willing to use the product.

In Web3, as with many new technologies, audiences don't trust what projects say about themselves; they trust what disinterested third parties say about them. This is the same as the logic behind the EEA launch. Audiences gain confidence when they see proof that others like them—or others they aspire to emulate—already use the product. What Chainlink did was flood crypto's social sphere with third-party confirmations, which are more trustworthy and therefore more valuable than anything a brand could say about itself on its owned channels. When journalists or potential partners search the project, they see a host of dapps and protocols using it, similar to how searches for Ethereum post-EEA yielded Microsoft and Santander.

The Chainlink strategy embodies our blended approach to discovery in Web3. Few of these new partner announcements are newsworthy enough to earn media coverage, even from industry outlets. Most of the partners are themselves small, early-stage Web3 projects. As marketers promoting these announcements, if we focused narrowly on the channel of earned media, our approach would fail. Instead, we acknowledge there is only a sea of perception that forms on multiple channels and we don't limit ourselves to any one channel. Owned blogs, and owned and partner social media, are perfectly reasonable channels to carry frequent, smaller-scale announcements. Publishing these consistently gives us the opportunity to reiterate our key messaging and reach new audiences. Although smaller announcements on owned channels don't usually yield step-function growth like the big moments, the Chainlink strategy helps marketers steadily add value to their projects in between them.

Events

Blockchain technology is the basis for systems that remove human elements from decision-making in favor of algorithms. But that doesn't stop the Web3 community from valuing human connection. For a movement that prides itself on automating trust and whose leaders spend their time in digital worlds and metaverses, Web3 is known to love IRL events. Although developers can inspect an open source code base themselves and decide whether or not to use it, they often discover new projects for the first time at meetups, hackathons, and conferences. Web3 communities are forming all over the world, not, as in the Web2 era, overly concentrated in hubs like Silicon Valley. That being said, the largest events take place in major global cities. If an event gains critical mass, it's common for community members to fly in even if they don't live nearby. Marketers of Web3 projects should identify relevant events and pitch their project leaders to speak at them.

At Serotonin we generally advise our clients against paying for speaking engagements. The best conferences prioritize offering audiences a high-quality experience. Those that charge all speakers, or "pay-to-play" events, are prioritizing something other than the audience experience. If the audience doesn't care about the content, it's less impactful for projects to speak anyway. Marketers should pitch for event speaking the same way they pitch journalists: by researching what event organizers might want from speakers and framing their pitch in accordance with what matters to them. If a conference won't accept a speaker without a fee, marketers should take this as a signal they need to further refine their pitch, or grow their project's brand recognition on other channels, to the point where conference organizers see it as a value-add for audiences to invite project leaders to speak. There are exceptions for later-stage projects with larger marketing budgets that specifically wish to target an event's audience. For them, paying to speak can be more economically efficient than paying to reach the same audience on other channels. In this case, projects should beware of their talk or panel being branded to attendees as sponsored content or relegated to a sponsors stage, which is guaranteed to have low attendance.

Marketers evaluating speaking opportunities for projects should pay attention to which stage a talk or panel is on (main stages are generally preferable over side stages), the time of day (the earliest, latest, and those during lunchtime can be poorly attended), and the quality and appropriateness of any co-panelists sharing the stage. When we prepare speakers for events, we always primarily consider the audience's perspective. For example, Web3 audiences often photograph interesting slides during presentations, so we encourage teams to include at least their brand name, if not website URL and Twitter handle, on each of their slides. Simple tweaks can be game-changers for discovery. An audience member scrolling through their photos at the end of the conference sees that brand name, or not. Every moment a brand can get its name in front of its target audience is an opportunity. In an IRL context, low-hanging opportunities for discovery abound, from including one's brand name on slides, to strategically placing a stack of branded stickers in the bathroom, to creating a branded scavenger hunt during which participants can scan QR codes on posters all over the city to receive tokenized prizes, to allowing guests to claim a POAP (proof of attendance protocol) NFT verifying they physically attended an event, adding them to a blockchain CRM for later. IRL events are an ideal platform for creativity.

Major events in Web3 that large numbers of people travel to include DevCon, the yearly Ethereum developers' conference hosted by the Ethereum Foundation; Ethereal, the curated event series I cofounded at ConsenSys that is now hosted by the industry media publication *Decrypt*; NFT.NYC, the largest yearly gathering of NFT enthusiasts in New York; the Miami Bitcoin Conference, where Vitalik Buterin first presented Ethereum in 2014; and Consensus, the long-running crypto conference hosted by top crypto news publication *CoinDesk*. Non-Web3-native events such as Web Summit, SXSW, the World Economic Forum meeting in Davos, the Consumer Electronics Show, Art Basel Miami, and the SALT Conference have grown significant Web3 content tracks and attract many Web3 community members. Web3 companies have started hosting their own conferences, including Mainnet, an event in New York hosted by crypto research firm Messari; SmartCon, hosted by Chainlink; and Permissionless, by financial media brand Blockworks.

Rather than organize full-scale conferences, and to make travel logistics easier on their communities, many Web3 projects opt to host side events concurrent with major conferences. The DeFi protocol Aave is known for hosting rAAVE, a series of nighttime rave parties after Web3 conferences, and many others host their own meetups, mixers, and happy hours. Projects focused on attracting developers host hackathons, staffed with mentors who help developers learn to build using their tools. Participants compete at specific challenges to win prizes, sometimes large financial rewards in tokens, other times investment capital. Many Web3 startups actually begin their lives as hackathon projects.

Full-Funnel Marketing

Most projects aren't optimizing events as a marketing channel. The reason is no surprise: in-person events are the meeting place for the Web3 community, and most of their participants are focused on socializing rather than driving ROI for their projects. Hanging out at conferences—going to talks, networking—produces serendipitous breakthroughs. I highly recommend it to newcomers and jobseekers. But for project founders, spending time at conferences comes with an opportunity cost because it's time away from coding, email, or managing teams. With so many tempting Web3 events happening at any one moment all around the world, it's possible for Web3 teams to FOMO (fear of missing out, can be verb or noun) into oblivion. After all, if you're constantly on the road, you're never building your product. So, for all things there is a season. There's a time for heads-down building and a time to hit the road to bootstrap demand. Before investing time and money on events as a discovery channel, projects should make sure to have their marketing funnel in place to ensure that their investment can be recouped as ROI.

It only makes sense to invest in top-of-funnel marketing activities if a funnel exists to convert discovery into engagement. For example, if founders travel to Paris for EthCC, the large yearly Ethereum community meetup, but they don't have a website live for their project with a link to sign up for the beta, join an email list, follow on social media, or join a community channel, then no matter how energetically they network or how well their presentation goes, the ROI will be limited. Potential users who come into top-of-funnel discovery immediately fall out of the funnel without being converted to take the next step. Marketers should approach events with a call to action after discovery and a metric target for success. Then they can measure the ROI to the project of the result, compare it to the time and money spent on the event, and use this data to make better investing decisions in the future. If a project doesn't know what it wants and hasn't set up a funnel to get it, it can be sure it won't. At the very earliest stage, projects are best served by lightweight market-testing activities that involve little investment of time or money. Events can be valuable for discovery, but they are most efficient as part of a larger funnel.

11

Engagement and Use

Start a Flywheel

FOR WEB3 PROJECTS, the steps of engaging users and then converting them to use the product are closely intertwined in the marketing process. This chapter describes how a potential user who has already discovered a project comes to engage with it more deeply, often by joining a Web3 community. Once this potential user has engaged, the project can retarget them with calls to action (CTAs) to convert them to using products and taking other useful steps to help grow the project. Engagement and use are a chicken and egg. Users who are excited about products often join their communities, eager to help out. Engaged community members reliably serve as beta testers and early adopters for products and features. This chapter also covers the importance of a project's early community to its long-term success and strategies for how to gather the right members.

Product Websites

To recap, the goal of discovery is to reach a target audience with a brand message. If the brand message is effective, it sends the potential

151

user into the next step of the marketing funnel: engagement. Discovery channels are all the channels where a potential user might first encounter a brand. The most common in Web3 are search, social media, earned media, and events. Then, depending on the channel where the potential user first discovers the brand message, engagement can take place on a variety of channels.

If the potential user is already motivated to find a solution to their problem and therefore discovers the product through Google Search, the next step toward engagement often looks like clicking through to the product website. Using Google Analytics and other tools, marketers can easily measure how frequently people searching for their keywords convert to website visits. The websites with the highest search rankings and descriptions that match what matters to target audiences will yield the most conversions.

If a user first discovers a project on social media, at an event, or through word-of-mouth, they may immediately search the project name and visit the website or look up the project on a social media channel. Social channels should be optimized to drive engagements, for example, like following the account, joining another community channel, and visiting the product website. More on this in the next sections. But many roads of discovery lead to the product website, which is the marketer's home turf for driving engagements. The most important engagements for marketers are those that enable the project to reach, or retarget, the potential user again, ideally leading to conversion to using the product. A project's website is the perfect place to accomplish this.

Wherever they came from, once the visitor arrives on the project website, they are greeted with a new set of signals. Unlike the discovery channel where the potential user first encountered the brand message and chose to engage with it, the website is completely under the control of the marketer. If the website doesn't work to convert the visitor to take the desired CTA, there is no faulting the algorithm, journalists who ignore the key message, or the noise in the background at an event. The product website is the marketer's playground, their owned storefront, to express the memetic package of words and images that is their brand exactly how they want, displaying their knife's-edge message exactly as designed.

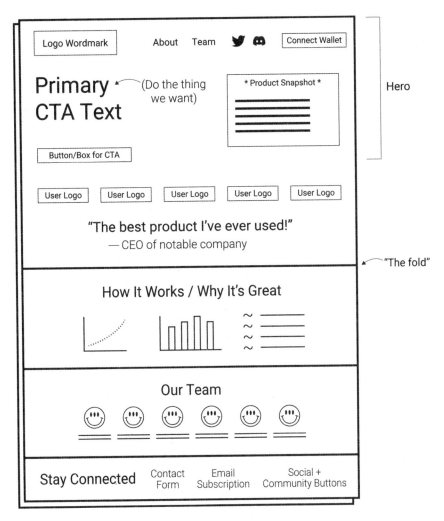

Figure 11.1 A basic Web3 product website

Figure 11.1 mocks up a basic layout for a Web3 product website, indicating placement of key design and marketing elements.

On a product website, the prime real estate is the *hero*, or what appears "above the fold" before a visitor needs to scroll. Analytics tools generally show that visitors are only willing to scroll so far. The further down the page, the more visitors have stopped scrolling, so the hero is the only section that all visitors see, assuming the website load time wasn't so long that they closed the tab. This happens, and marketers should assiduously work to avoid it; most visitors are ruthless and will only wait a fraction of a second for a page to load.

Each element on a product website should occupy this prime real estate directly in proportion to the importance of the CTA to the success of the project. Ideally, marketers should identify a single primary CTA, and this should occupy the largest and most important position on the hero. Some examples of Web3 projects' primary CTAs to convert to use are "Sign up for the beta," "Register for the whitelist," "Download x," or "Join the DAO." The CTA is often near the key brand message. Often there is a product screenshot to the right or left of the primary CTA, which ideally signals to potential users that the product has an easy UX, increasing the chances of conversion into use. Marketers can use analytics tools to A/B test website copy and images to maximize the chances of conversion while being careful not to implement language—even if it converts the best—that doesn't tightly match the primary value proposition of the product; otherwise, users may convert to use originally but then fall out of the funnel at the stage of retention, disappointed that the product doesn't solve their problem.

Although the primary CTA should get the most prominent placement on a product website, the hero can also feature other, secondary CTAs. On the top right often are buttons that link to Twitter, Discord, or any other social and community channels the project manages. Dapps often place buttons for users to connect their Web3 wallets in the top right-hand corner. On the top-left or center is usually a navigation bar with headings such as "Team," "Blog," "About" or "How it works" and sometimes the name of a token or product feature with more information about it. Sometimes these are connected to dropdown menus. To optimize for discoverability on search, these headings can be named after the most frequent search terms for the project. Some projects also have sign-up boxes for email newsletters. For projects that haven't launched yet, an email sign-up box can be the primary CTA, allowing the project to retarget potential users when a product beta or token becomes available. Marketers should make sure to include key CTAs in the hero section. Less important CTAs, along with more detailed information for more motivated visitors who want to learn more, can occupy lower sections of the page that require the visitor to scroll.

Web3 marketers may or may not choose to implement Google and Facebook tracking pixels on their websites. When our department at ConsenSys first met with the team at Truffle (the leading Ethereum development framework within the ConsenSys mesh), we presented a number of ideas to help Truffle better understand its audience and how well its website was working. Truffle was excited about our presentation but hesitant to add tracking pixels to their website, a step that seemed obvious to members of our team trained in Web2 best practices. Truffle was ideologically opposed to the Web2 practice of surveilling users and refused to add the pixels, even if they could help increase their user base faster, which was their primary business goal. We didn't push back. Instead, our marketing team learned something about our audience that day. The developers who were Truffle's target audience disliked the practice of collecting data on users. Any website visitor can easily click "inspect element" or "view source" to identify a tracking pixel in the page's HTML. Although probably only a small minority of visitors would actually refuse to use the product due to the pixels, or worse, vocally complain on social media, at that point in the history of Web3, those ideologically driven technologists were our key target audience, those most likely to become Truffle power users. The brand risk of implementing a pixel wasn't worth the potential benefits to user growth.

Though Truffle may or may not use tracking pixels on its website today, we still periodically encounter teams who wish to refrain from using Web2 marketing technologies, especially for user data collection. For example, at Serotonin we worked with Orchid, a Web3 VPN. As a privacy tool, they wished to target audiences concerned with online privacy, who are also more likely to inspect a page's HTML. Implementing tracking pixels was out of the question; we would develop novel methods for monitoring our marketing funnel. Sometimes this means finding alternative A/B testing and analytics tools, such as Matomo, which offer some data while respecting user privacy. But the lesson broadly applies to Web3 marketers who must be flexible, pay attention to audience preferences, and not overly depend on practices from Web2.

A good product website evolves alongside the product. We recommend Web3 projects publish a simple splash website as soon as they have determined their visual and verbal branding, knife's-edge key message, and primary CTA. A prelaunch project's primary CTA catches the potential users who discover the project on various channels and are willing to engage, in a holding tank, allowing the project to target them again later when it's ready to convert them into use. Without a CTA in place to act as a holding tank, projects may drive discovery through channels such as social media, search, at events, or through word-of-mouth, but all these activities will be inefficient at producing ROI for the project because the energy can't be captured anywhere. Without driving toward a backstop that can capture interest, usually a product website with a CTA, top-of-funnel marketing activities are basically useless.

Between launching their website and social channels and actually launching their products, projects should concentrate on growing their owned audiences on social media and community channels and email newsletters. They should use these channels to keep their engaged communities up to date on progress with their products, sharing information about their teams and the technology they are building, including any announcements, events, partnerships, funding rounds, or asks for the community to get involved, such as with community management or a bug bounty program. Once a product is live, the project's website needs to change its primary purpose from funneling engaged potential users into holding tanks to be reached later for conversion into use, to converting visitors directly into using the product. It's time for the website's primary CTA to shift.

Marketers designing product websites should keep in mind that they're still fighting against the forces of indifference, inertia, and skepticism at the engagement stage of the marketing funnel. Just because a potential user has arrived at the website doesn't mean they will convert. The content of the product website needs to convince them to answer the call to action. Potential users are always wary of new, less-established projects, especially in a sensitive space such as Web3, where connecting one's wallet to the wrong dapp or sending crypto to the wrong address can result in hacks or value being permanently lost. The content of the website must include knife's-edge messaging to dispel indifference by convincing the potential user they really have a problem, and inertia by demonstrating the product is the best solution. It must also assuage skepticism by reassuring the potential user that others like them are using the product; that the product is endorsed by trusted third parties like notable investors, partners, and code auditors; and that it's been proven over time with a history of use.

The "wall of trophies" and "instant incumbency" strategies that follow have worked for Web3 projects. They may or may not work for you. Whether you decide to use or adapt these—or develop your own—should depend on first knowing your audience.

Wall of Trophies

The wall of trophies tactic pairs well with the Chainlink strategy. As early as possible, projects should strive to attract users who resemble the users in the broader audience they want to target. They should broadcast these partnerships using the Chainlink strategy, harnessing them to drive top-of-funnel discovery on search and social, and also showcase their partners' logos prominently on their websites, sometimes even in the precious hero section near the primary CTA. Website visitors who see the names of prominent users—perhaps companies just like theirs, or larger and more notable ones, or well-known individuals who use the product—will be less skeptical and more likely to take the CTA.

If no comparable companies or individuals use a product, a potential user takes a longer path toward use stymied by more inertia, because they need to either do research for themselves or, if they represent a team, generate a detailed argument to convince colleagues that the product is right for their business. A prominent wall of trophies enables many potential users to skip this time-consuming step and convert directly into use or another CTA. If a project sees that its competitors are already using a tool, this can be even more compelling, because it doesn't want to miss out on an advantage.

Walls of trophies can take many different shapes. For some projects, the trophy to display isn't partnerships with other companies, it's metrics like total value locked (TVL). The mechanism is the same. A DeFi protocol with high TVL demonstrates to a new potential user that other DeFi users just like them are willing to deploy their funds in this protocol. For this reason, DeFi dapps and liquidity pools often prominently showcase these types of metrics near their primary CTAs.

The EEA website we launched in February 2017 was a simple splash page with a logo, key message, and a social button to follow on Twitter. The most prominent image in the hero was a wall of logos that included

Microsoft, Intel, JPMorgan Chase, Accenture, and more. The primary CTA was an email capture bar to submit a request to join the EEA. Many companies large and small filled out the form, and soon the EEA had a long list of qualified leads waiting to join the organization.

Projects of all sorts and sizes shouldn't hide their users under a bushel. Instead, they should feature walls of trophies on their websites, the logos of their most valued partners, cutting through blocks of skepticism, indifference, and inertia to drive conversion.

Instant Incumbency

Because of the risks inherent with Web3 dapps, potential users are often wary of new projects and companies. It doesn't help if their teams are populated by pseudonymous Twitter accounts, especially if the accounts have few followers. In combination with the wall of trophies and Chainlink strategies, new projects should consider the approach to their brands that we call instant incumbency. If a project is competing with larger, more established incumbents or institutions, it helps to look legitimate. That means adopting the visual and verbal style of more established competitors into their brand look and feel, perhaps with a modern touch unique to the project's new brand that helps it stand out against the background of what the audience expects. A brand that feels like an incumbent overnight isn't an exact copy of another brand; it needs its own twist, but it can take inspiration.

Marketers should pay attention to any channels or interaction points between potential users and the brand that consciously or unconsciously signal that a brand is small and new. For example, when I joined ConsenSys, which was at the time small and new, our website, blog content, and even automatic signatures on employees' email included detailed descriptions of what ConsenSys was. This sent the message to potential audience members that we didn't expect people to already know about ConsenSys. With all we were doing to build Web3, our team knew ConsenSys was positioned to become a significant technology player. We would get there faster if potential partners, especially at large institutions, saw us as a more mature and established company.

Our team removed any language that suggested to counterparties that we assumed they weren't familiar with us. We still featured an

"About" page on our website where visitors could read about ConsenSys, and we included the standard blurb at the bottom of a press release with language about the company; our goal wasn't to prevent new visitors from learning about ConsenSys. But we consciously stopped presenting ourselves as a startup that people probably didn't know. Once we scrubbed these explanatory lines from all the channels where they existed, when new target audience members discovered our brand for the first time and didn't recognize it, they assumed the error was on them. I think they wondered, "How could I not have heard of ConsenSys?"

Some Web3 projects, such as those that emerged during DeFi foodcoin summer, intentionally brand themselves as being new and exploratory. Their websites featured a 1990s vaporwave aesthetic, cartoons, and cat memes. These brands were perfect for the mostly young, often technically proficient risk-takers who identified themselves as degens. Here the "instant incumbency" mindset wouldn't have worked. Every project should construct its brand and key assets like its website out of a deep understanding of their target audience and what matters most to them. At Serotonin, our process for working with a new project always starts with building a visual and verbal brand, the foundation of any further marketing efforts, based on our understanding of their target audience. A high-quality brand, defined as matching its audience, increases the likelihood that engagements will convert into use.

Fostering Community

After a new potential user discovers a project, they may decide to engage with it by visiting its website. Alternatively, they may visit its social media account. When the new potential user arrives on the social channel—let's say a Web3 project's Twitter page—without necessarily consciously thinking about it, they perceive from the size of the project's following whether it's large and established, with many community members, or whether it's small, new, or unable to attract a following. Pushing against the headwinds of indifference, skepticism, and inertia, they may decide to follow the account if they think the content is interesting or want to hear updates from the project. The new potential user might click through the links in the Twitter bio and arrive at the project's website, a collection page on an NFT marketplace like OpenSea, or another location where they could be converted via CTA to make a purchase or use a product.

They also might click links in the Twitter bio that send them to the project's *community channels*. These are simply communications platforms where members of a Web3 community may gather, not to be conflated with the overarching idea of community in Web3, which refers more abstractly to the group of people connected with a Web3 project, who may or may not have joined a particular platform. For most Web3 communities, these platforms are Discord and Telegram. Slack, WhatsApp, and others are used also, but less frequently. The reason Discord is so popular with the Web3 community is that it was early to implement plugins that connect with Web3 wallets. This enables Web3 projects to spin up Discord channels and make them available only to holders of a particular fungible token or NFT or a certain quantity of that token. For many of the most popular NFT collections, these token-gated communities are the invite-only members clubs of the internet, each behind a different velvet rope, with its own custom criteria for entry. Inside them, members access updates from project teams, exchange information and ideas, get to know and trust each other, organize events and meetups, contribute to projects, and form new ones of their own.

The concept of community in Web3 can sound annoyingly high-minded or abstract to outsiders, but it actually means something specific. It refers to the single category that forms when a Web3 project collapses those of investor, builder, and user into a single economically aligned group. Participants in this group contribute to the project fluidly and in self-defined ways, sometimes contributing to its code base, answering a challenge or participating in a bug bounty for a reward, and sometimes deploying capital into the project. Their incentives can run the full gamut from extrinsic to intrinsic. Some communities are primarily social groups that offer participants a sense of identity and shared purpose.

For example, Friends With Benefits is a token-gated community on Discord that requires 75 FWB fungible tokens in a Web3 wallet to join. Its members, generally Web3 enthusiasts in the cultural sphere of arts and entertainment, organize parties and get-togethers, share new projects and job opportunities, or exchange "alpha"—scoops about trading tokens or NFTs. A different example: the well-known NFT project Bored Ape Yacht Club has a Discord server for its community. Some channels on their Discord are open to the public; others are token-gated and visible only to holders that connect their Web3 wallets and verify that they own a Bored or Mutant Ape. Similarly, these members are sharing insights and answering each other's questions. They gain special early access to information about the project and are in direct contact with its founders and leadership.

Indeed, in the largest and most established communities, project founders are actively engaged, making themselves visible and available. Even celebrities launching NFT collections should make themselves regularly available to their communities. Authentic and regular access is one of the perks Web3 audiences expect. Serotonin worked with Pace Gallery, one of the largest art galleries, to launch a Discord community for their NFT platform just as it was about to launch an NFT series by Jeff Koons, one of the world's most famous living artists. To their surprise, we insisted that Koons himself do an AMA ("ask me anything" session) on Discord and engage directly with the community there. This project also shows that access, although important, doesn't have to be extreme: Jeff Koons didn't have to be available on Pace's Discord around the clock, but his AMA engaging directly with the community there—on their home turf, as it were—demonstrated his (and Pace's) sincere commitment and interest in Web3.

Considering what these communities offer, the notion of community in Web3 is more profound than it originally appears. Our atomized, individualistic culture can be painfully lonely. Many of us felt this way especially during the pandemic. As we enter a world in which our physical reality is more constantly mediated by and enmeshed with digital reality, it's easy to assume that we'll be driven further apart. However, this isn't necessarily true—Web3 offers us new structures to come together. These communities are an early example of a new type of structure, a unit for both group sense-making and economic survival.

The formation of Web3 communities is philosophically fascinating. Almost every technology that has made the most impact on humanity has allowed people to collaborate in larger groups, and Web3 is no exception; Web3 communities and DAOs demonstrate this point. But when it comes down to actually managing Web3 communities on platforms such as Discord and Telegram, we come back down to earth. These communities require granular day-to-day operations to get off the ground and then continue to thrive and grow. Communities on Discord meet regularly on weekly or monthly phone calls, during which project leads usually update the group on progress and road map. Most DAOs have their own communities on Discord whose members are the members of the DAO. These DAO communities participate in similar activities, but they also use the channels to discuss their votes on investing opportunities and new members.

In the context of the marketing funnel, the CTA to join the community usually appears at the engagement stage: it can be a primary or secondary CTA on a product website or a CTA on a social media account. On the shallowest level, community channels, like social media channels, function as holding tanks for potential users who are willing to engage, enabling projects to continually retarget them with CTAs to convert to using a product. Although this framing makes sense in a Web2 context, it's far too reductive a way to understand how communities function in Web3. The community isn't a group waiting to be marketed to by the project; they become part of the project themselves. In Web3, the marketer merges with the community, the community with the marketer. We all become one.

But nirvana isn't achieved overnight. Before launching a community channel, projects should make sure they are configured and staffed properly. The day-to-day of Discord channels is often run by pseudonymous mods (moderators) who can be anyone, at any level of seniority. Especially at first, team members should plan to make themselves available to be mods. Third-party community moderation services exist, and they are better than nothing, but really, we tend to recommend the team behind the project be the ones moderating the community. They can hire teammates from firms with this experience, but from the perspective of the newly forming community, it should be run by its own team. As the community grows, project teams sometimes offer the most engaged community members incentives to begin serving as mods themselves. This is especially helpful to projects that attract global communities and therefore look to secure 24/7 coverage.

It's important for projects to let communities do their own thing. Members will set up their own subchannels and launch their own conversations. Marketers should pay careful attention to see what their communities do organically that works in service of the project's success. Then their job is to fan those flames, as in our examples of

Crypto Coven's "Web2 me vs. Web3 me" campaign and Moonbirds supporting the emerging Ladybirds subcommunity. These projects noticed their communities were doing something helpful and, instead of clamping down and insisting all ideas need to come from project headquarters, offered these self-organizers the resources and authority to grow the movement in their own direction. This is exactly what we did early on at ConsenSys, organizing a network of Ethereum meetup leaders and providing them with content and resources.

On the flipside, it's important not to let things get too out of hand. It's contingent on community mods to clearly set the rules— usually on the #general or #welcome channel that users first see—and enforce them, sometimes even banning members. If a community channel becomes all speculators talking about price or starts violating community standards for behavior with inappropriate content, it becomes a place where target audience members don't want to live.

The Community Spaceship

It matters a great deal who a community's first members are. They disproportionately set the tone for the rest of the community that forms. Once a community grows to a certain size, the original project team effectively loses control over the tone of the community and whether their target audiences will want to join it. The early days when they are in control is the time the project team can make the most impact by curating members. A good analogy for starting a Web3 community is an alien arriving on a spaceship, sending down a ladder somewhere, having humans climb the ladder and board the spaceship, then flying back off into space. The aliens are then stuck with those humans on a spaceship forever. Where should the aliens let down their ladder? What kinds of humans do they want on their spaceship? Given the long-term consequences, the aliens should consider carefully before arriving on earth. The stakes are high: getting the right community members is far more important than community size at the start.

Most projects' ideal community members aren't just in it for rewards; they can be interested in the technology generally, using the product, contributing toward a larger mission or set of values, or connecting with others. Some communities, however, are explicitly economically driven, such as the DAOs that make governance decisions for a DeFi protocol. It can be essential in this context that community members practice rational self-interest and discuss tokenomics and math. The tone and content of the MakerDAO Discord server will be different from that of Bored Ape Yacht Club's. If a project has a token or NFT associated with it, all the holders of that token are generally considered community members, and many elect to join community channels like Discord to connect with people with similar interests, stay informed about the state of their assets, and contribute where they can to the project's success.

Who holds a project's token and their incentives to buy and sell the token can impact short- and long-term token price. If a project distributes tokens to people who don't care about the project, those people will most likely sell them, and without sufficient buy pressure, the token price will collapse. However, if a project distributes a token by a mechanism such as retroactive distribution proportional to community members' history of using the product, those members are more likely to retain the tokens, join and participate in community channels, and continue using the product.

Where Web3 projects find their earliest community members and token holders can determine the fate of their projects. For most projects, the ideal is growth over time as the project builds and executes its road map, adding on new fundamentals and markers of success. Airdropping tokens to anyone who performs a certain action can generate large communities or social followings overnight, but this and other, similar short-term tactics can lead to long-term struggles with community tone and token price. Similarly, projects that buy followers or community members can expect them to be mercenary. The best way for Web3 marketers to attract community members is to clearly present the value proposition of their products and communities with knife's-edge messaging, ensure that messaging gets discovered on top-of-funnel channels, and deliver against their promises.

At Serotonin, we usually suggest Web3 projects form communities before they launch a token or NFT collection. That way, there is already pent-up demand for the product. Bootstrapping a community before launching a token gives project founders the opportunity to test whether target audience members are receptive to their value proposition, and to tweak that value proposition until they find a fit with an audience. They should continue to iterate their offering until they're sure that their community will be receptive to it, never hesitating to ask their community members directly, Would you be interested in *x*? or, What would you change about *y*? Projects can learn a great deal from their early communities to increase the chances that launches and drops achieve their goals.

We also often recommend that new projects offer their early communities value in a Web3 context before asking them to pay for something. This helps build trust and attract the types of community members who are interested in the actual project. One example of perfect execution of this strategy is the NFT influencer Gmoney, who whitelisted the wallet addresses of people who had attended certain conferences based on the POAP tokens in their wallets (proof-of-attendance protocol, a protocol that generates tokens that prove someone has attended an in-person event), and allowed those wallets to claim an Admit One pass, an NFT that enabled them to access his token-gated community channel on Discord. A *whitelist*, sometimes called an *allow list*, is a pre-vetted list of public wallet addresses that are allowed to perform a particular action on-chain, such as claiming a claimable. Admit One community members on the Discord server received direct access to Gmoney and information about the project. Because of Gmoney's reputation for good taste in NFTs and supporting NFT communities, the Admit One NFTs quickly became valuable, sometimes priced as high as $15,000, even though he originally gave them away for free. It was only after building this goodwill on top of his strong existing reputation that Gmoney launched his NFT fashion line, 9dccxyz, whitelisting Admit One NFT holders for early access to purchase items from his collection.

Building community first, and giving away items for free to start, may sound overly generous or like too much prework before starting to capture revenue—at least to those with a Web2 mindset looking to round up "consumers" and charge them money. There is, however, a strong logic behind the practice. Even if a Web3 project doesn't immediately profit financially from its first community formation efforts or its first token or NFT launches, it has still performed an economically valuable activity by establishing an on-chain history of wallet interactions with the project. These wallets can be retargeted indefinitely and become the marketer's dream for customer relationship management (CRM—for more on which, see Chapter 12 on retention) and over time, put the project in a position to perform a retroactive distribution to token holders already proven to like using the product, people who would be more likely to buy or hold than sell. Though early Web3 community-building strategies can sound counterintuitive at first, marketers should keep in mind the North Star of building a healthy community that grows over time.

Content

Marketers must be able to furnish their chosen channels with content. On social media platforms such as Twitter and Instagram, that can mean custom images. On YouTube and TikTok, that means videos. Before marketers spin up a particular channel, they should make sure they have engaged the correct staff, either internally or through agencies or partnerships, to ensure the channel stays active with a regular cadence of publishing new content. Nothing looks worse to a visitor than an empty channel with no posts, engagements, or followers, or one that posts inconsistently with an irregular voice and message.

Marketers shouldn't launch channels until they have the proper staffing in place and a future-looking content calendar so the team always knows what content will be published around the corner. The marketing team should set objectives and key results (OKRs) for content performance, not focused on publishing a certain amount of content per time period, but on the outcomes produced by the content (for greater detail, see Chapter 8, "Set Key Metrics"). A few examples of outcomes content can produce are: an $x\%$ increase in website traffic, y number of social media engagements on a post, or driving z number of qualified leads into an email capture. The content itself, whether it's in the form of social posts, blogs, op-eds, industry reports, or event presentations, should be closely tailored to fit the interests of its target audience and reiterate key brand messaging. Reiterating this messaging in blog posts helps websites gain domain authority and rank higher on search, even if not every post drives a massive amount of traffic. Projects should ensure that they publish blog content with each of their main keywords in the title to increase the chances that they rank for those terms (for more on blogs and search ranking, see the "Search" section in Chapter 12). The more relevant the content is to its target audience, the more they will open and share, compounding the benefits of discovery.

Once target audience members sign up for an email list, join a community channel, or follow a social media account, the project is in a position to continuously reengage them with content, including CTAs to start using a product or try out a new feature. Social media platforms are mostly algorithmic, which means projects can't reach their full audiences at will with new content. On a community channel or email listserv, there's less intermediation, and marketers can decide when they want to share content, as long as audiences will pay attention. We typically recommend, if projects have the staff to create and distribute emails, and if the channel makes sense for their audience and business goals, that they spin up email marketing. Discord or Telegram could suddenly disappear. The Twitter algorithm could change one day to privilege certain types of accounts over others. And certainly, MailChimp and other email platforms could have problems, but there's nothing more dependable than a spreadsheet with the addresses of audience members who have agreed to receive emails. Email is one of the few channels a project can truly own, and it can become an excellent way to distribute content.

Content and email newsletters don't always need to convert users to using a specific product; they can be helpful for building a store of goodwill that can later be used toward converting users. Early on at ConsenSys, we didn't always have a product or token launching, but we realized the burgeoning Web3 community needed a regular source of updates about everything being built on Ethereum. Our marketing team began writing a regular newsletter, and our email list became a destination for anyone curious to learn about the Ethereum ecosystem, not just those who wanted to receive product marketing messages from ConsenSys. When we did have a product to market, we were able to reach a larger and more engaged audience because of all the time we'd spent building credibility and adding editorial value in the past.

Content can be a godsend for projects that haven't yet launched their products. When projects come to us at Serotonin, they often want to start building social followings and communities, but they aren't sure what to say to them without having a product actually live yet. The answer is content. Prelaunch projects should think of themselves as niche content brands. A Web3 privacy project can spin up a blog series and newsletter about Web3 and privacy. A prelaunch DeFi protocol can create educational content about how newcomers can start using DeFi. A woman-led NFT collection can publish lists and analyses of the best women-focused NFT communities or create a network of other women-led NFT projects and cross-amplify each other's content on social media. If projects generate demand through content and concentrate it on specific channels, when they *are* ready to launch products, they need simply to open the floodgates. Content-focused marketing activities can and should start long before launch.

12

Retention

Build a Moat

MARKETERS BATTLE THE forces of indifference, skepticism, and inertia at every stage of the marketing funnel until they reach retention. Retention is the step when the potential user has already converted to becoming an actual user. The goal of retention is defense: to keep the user coming back for more and avoid losing them to a competing product. At this point inertia actually works in a project's favor. Just as in Newtonian physics, when an object in motion stays in motion, the user of a product will continue using it, so long as they continue having the same problem and the product continues to offer the best solution to their problem. A few of the best mechanisms to retain users are having a sticky product, offering incentives, and building a reputation for keeping promises to the community and delivering against a road map.

A *sticky* product is defined as one to which users are likely to return and continue using and engaging on a regular basis. The opposite of a sticky product—let's call it a slippery product—either solves the user's problem after a limited number of uses and sends them on their way, or lacks a built-in mechanism to motivate a user to return. A Web3 wallet such as MetaMask is an example of a sticky product, because moving all of one's tokens into a new wallet causes friction, and users must be motivated in order to overcome friction. The more motivated they are and the less friction required for a user to switch products, the more likely they are to slip out of using the original product.

DeFi protocols can be less sticky products. If another protocol offers higher yield per perceived amount of risk, a user is likely to go else-where. That being said, DeFi protocols, similar to other Web3 projects, can use community channels to make their products stickier. NFT collections frequently increase their stickiness with their community channels, which can become part of members' daily lives. If a Bored Ape NFT buyer joins the Bored Ape Yacht Club (BAYC) Discord server, and the conversation on those channels is entertaining or useful, that buyer is more likely to retain their Bored Apes or buy more BAYC NFTs to continue having access to the community channels—if they sell all their BAYC NFTs, the token-gating mechanism on Discord kicks them out of the server, and they lose access.

Another mechanism NFT collections use to retain community members is rewarding NFT holders with further NFTs. For example, BAYC airdropped Mutant Ape Serum NFTs to all the Bored Ape NFT holders. Holding a Serum NFT allowed the Bored Ape NFT holder to mint a Mutant Ape. Mutant Apes became valuable NFTs of their own, though currently less valuable than Bored Apes. Holders also traded the Serum NFTs themselves. NFT projects like BAYC that offer existing holders additional fungible tokens or NFTs develop a reputation over time for rewarding their communities, which leads holders to retain or buy more of them.

Projects can also offer existing holders—or any set of on-chain wallet addresses they want, even those with a history of using a different product, because the history of wallets that have engaged with a given dapp is public on the blockchain—the opportunity to join a *whitelist* or *allow list*, which entitles them to future airdrops, claimables, or opportunities to purchase further tokens. When Adidas and Prada launched a digital artwork NFT drop together, they reserved 1,000 of the 3,000 NFTs to be minted for wallets that had purchased a previous collection of Adidas NFTs.[1] This is the future of reward programs. Any Web3 project—from DeFi protocols to Web3 utilities to NFT collections, regardless of whether their product is inherently sticky—can use the substrate of Web3 to design incentive systems that improve retention.

Web3 incentives to promote retention can be intrinsic or extrinsic. Take Friends With Benefits, for example. It is a token-gated DAO community on Discord that is a meeting place for individuals joining the Web3 movement from the cultural sphere. Many of its members work in music, arts, entertainment, media, or gaming. Some have spent years working in Web3; others are looking for information, seeking jobs, or new product recommendations. Friends With Benefits frequently hosts parties during major Web3 global conferences that are open to community members. The community has a fungible token called FWB that can be purchased on DEXes such as Uniswap. Although anyone can buy the token and participate in the DAO's governance, only approved applicants can become proper "members" who are invited to tiered participation contingent on holding FWB tokens. An approved member holding five FWB tokens can connect their Web3 wallet and gain admission to "local" events and limited Discord channels; a "global" member holding 75 FWB tokens has full access to events and all Discord channels.[2] At the door of IRL events, attendees may be asked to show mobile wallets that prove they hold a minimum number of tokens. Most of the members of Friends With Benefits don't hold FWB primarily because they speculate the value of the tokens will increase—though through the logic of the tokenomic design, the more demand to join Friends With Benefits, the more buy pressure on the FWB tokens required for membership, and therefore the higher the price—but rather because they wish to enjoy the benefits of the community. If they were to sell their tokens, the Discord server would boot them out, and they would lose access to in-person events and the digital community. To the degree that projects deliver their communities value that isn't purely financial, they can promote retention with intrinsic incentives. Any community—from a DeFi protocol to a Web3 utility to an NFT collection—can offer intrinsic incentives to retain users or holders by fostering active community channels with content that engages their target audience.

A well-known example of a community forming around the potential for extrinsic rewards with the result of promoting retention is the Link Marines. Chainlink, the decentralized oracle project, developed a hardcore group of fans that called itself the Link Marines, after the Chainlink token, LINK. At the start, the Link Marines mostly appeared to be young, predominantly male crypto investors who spent time on the online message board 4chan. In their community channels, the Link Marines created their own memes, often politically incorrect, about the strengths of the Chainlink project, which they shared rampantly on Crypto Twitter. They dared each other to buy more LINK—even to the point of holding a "suicide stack" of the token (that is, so much that if its price crashed, they would lose most of their value)—and berated anyone threatening to sell their LINK. The Link Marines community wasn't for everyone, but its members had fun and found companionship in coming together as a group to try to achieve economic goals. As the group achieved critical mass, it arguably contributed to the price of LINK soaring over many consecutive months. Even token investors who weren't explicitly part of Link Marines would hear or see on Crypto Twitter that Chainlink had an army of militant fans. "It must be good if it's inspired so much confidence from all these people," they might think. This third-party confirmation probably made some new potential buyers more likely to participate and existing holders of LINK more likely to retain their tokens instead of selling them.

If projects forming their early communities are like spaceships lowering their ladders for people to board, as per our previous analogy, who boards those ships matters a great deal for retention. Many Web3 projects appear at first glance to have thriving communities because they have large followings on Crypto Twitter. However, if projects attracted those audiences primarily with giveaways, not caring if the early audience members are likely to actually use a product or contribute to the project, the resulting communities are unlikely to stick around once the rewards run dry.

Those who have spent time in Web3 communities recognize the exhausting refrain of "wen token," a partial joke that members use to ask repeatedly when a project will launch its token, often suggesting they will leave once they've collected and dumped it. Distributing fungible tokens or NFTs to existing users of a product is a wise strategy to optimize the chances those audience members will choose to retain their tokens and continue using the product. Past behavior is the best predictor of future behavior. If a project airdrops its tokens to wallet owners who don't care about the project, they will most likely sell them as soon as possible, cashing them in for tokens they believe will increase more in value, or for stablecoins or fiat currency. The highest-leverage mechanism a project can use to promote long-term retention is carefully considering how it forms its early community. Marketers should thoughtfully select their target audiences, welcoming in those who are most likely to buy, use, or contribute.

Oftentimes communities stick around because they believe in a project's road map. The Azuki NFT collection by gaming project Chiru Labs is a perfect example of retention as a function of winning—and losing—community trust. Azuki NFTs feature anime-style avatars with different characteristics. Beyond PFP utility, each NFT grants the holder access to The Garden, Azuki's token-gated "metaverse" community. Although Azuki's website straightforwardly defines the current state of that metaverse as a Discord server and Twitter Spaces, the website also includes a robust road map of future activities and capabilities for the collection—IRL events, exclusive apparel drops, an immersive metaverse world, a variety of games and experiments related to the characters in the collection—many of which point to possible events that could drive step-function growth in the value of the project (for example, a land drop accompanying the metaverse world launch). The beautiful design of the NFT art, the website, and across all of Azuki's social and community channels inspires confidence that the team behind the project could execute against its full vision. An Azuki NFT buyer might retain the token for intrinsic or extrinsic reasons, or a mix of both. The potential for growth is exciting to speculators looking to resell the token later at a higher price; the idea of accessing special channels, events, and someday a game and metaverse world, is intrinsically exciting to others.[3]

Whereas Azuki's road map inspired confidence—the January 2022 launch sold out in three minutes, followed by a steep rise in prices on the secondary market—a May 2022 blog post by one of the founders discussing his abandonment of multiple previous, failed NFT projects led to a massive loss of confidence, especially when Crypto Twitter investigators pointed to potentially shady practices at these past projects. Token prices tanked and as of this writing have stayed well below their early 2022 peak. Azuki continues, but whether it can regain the community's trust, retain members, and build out its initially promising vision remains to be seen.[4]

Often newcomers to Web3, especially to the NFT space, criticize the industry as being full of speculators, with few actual users. These skeptics often don't understand Web3 and approach it with a narrow Web2 mindset, where builder, user, and investor are separate roles, each with at least partially adversarial incentives toward the others. With this mindset, a community full of speculators means a project lacks real users. What these skeptics fail to understand is the way Web3 projects collapse the categories of builder, user, and investor into a single aligned unit called community. The motives of community members are often blended. A community member may behave like a speculator and resell a token for more value; they may also decide a game is so much fun to play, or a community so valuable to participate in, that they continue to hold the token. Some community members join with a particular agenda; others watch how a project executes against its road map, rewards loyalty and use, and fosters a sense of belonging on its channels to determine their behavior.

To retain users or holders, a Web3 project should plan its long-term technology road map, consider publishing it on its website, and keep existing community members up to date on its rollout with regular messaging, community meetings and forums, and public social posts. On the one hand, projects that follow through on their road maps over time inspire trust and earn retention. On the other hand, projects that deceive their communities or fail to deliver on their road maps can not only harm their own reputations but also chip away at the credibility of the Web3 space at large. Unfortunately, a number of scammers lured by the potential for value capture, as well as incompetent teams that overpromise and underdeliver, have harmed the reputation of the early Web3 industry, especially in the eyes of generalist users, institutional investors, the media, and some regulators. However, as more high-quality Web3 projects emerge, making plain to these audiences the beneficial use cases of the technology, and as the Web3 community evolves an immune system for identifying scam projects—its own decentralized version of *Consumer Reports* or the Better Business Bureau, which we are already seeing with companies such as Messari that independently evaluate Web3 projects—it will naturally become easier for potential users to determine quality. Scams will always exist wherever there is money to be made, but many will be deterred over time as that immune system strengthens, similar to what we saw in the early web.

Web3 marketers are only just beginning to unlock its potential to retarget and retain communities. Increasingly, they are realizing that the blockchain is the best CRM (customer relationship management) software known to humankind. In Web2, what platforms like Facebook and Google are selling to advertisers is largely identity information that enables the advertisers to granularly target potential customers with particular attributes that make them more likely to buy the product being advertised. These platforms collect identity data from users and use correlations from this data to extrapolate buying behaviors. For example, a man living in a rural farming area is more likely to buy a tractor; a woman with a cat as her profile picture is likely to buy cat food. Possibly, though, the man lives in the countryside because he likes hiking, and the woman is actually allergic to cats. This example is obviously oversimplified, but it shows the limitations of extrapolating buying behavior from identity. The best way to predict future buying behavior is simply to see past buying behavior.

On the blockchain, there is an immutable public record of the wallet addresses that have purchased any on-chain assets. These addresses may or may not be linked to any discoverable identity information—but such information is unnecessary. If a marketer for a Prada NFT knows which wallets have previously purchased Prada NFTs—or Louis Vuitton NFTs, for that matter—they can target those wallets with a whitelist, airdrop, or claimable. They don't need to know anything about user identity because they have direct access to the past history of buying behavior. By using the blockchain, companies can achieve their goal of targeting specific audiences, while preserving users' pseudonymous privacy.

Marketers can get even more surgical using on-chain data for retargeting or retention. For example, let's say Prada wants to incentivize its community to hold on to NFTs as long as possible. In this case, marketers could target wallets that have held a given NFT for a minimum period of time. If Prada wanted to reengage community members who sold their NFTs, they could target wallets that at one point held Prada NFTs but didn't any longer. If marketers want to take respecting user privacy a step further, or if users demand it, they will increasingly be able to leverage zero-knowledge proofs (ZKPs), a technical advance in the Web3 space that enables verification that a wallet holds certain tokenized assets without visibility into the contents of the wallet. Blockchain platforms such as Ethereum and Polygon are actively integrating ZKPs in order to unlock this functionality. With pseudonymous wallets plus ZK integrations, Web3 offers marketers truly powerful CRM tools to create custom campaigns to incentivize desired behaviors, without knowing the identity of a particular user or even what other property they hold.

Just as Web3 opens the door to new retention strategies, it can also make products more slippery. Take the example of SushiSwap and Uniswap from Chapter 3. Because Uniswap is open source like many Web3 projects, it was trivial for another project to copy its code, relaunch it with minor modifications, and quickly stand up a competitor. This is the beauty of open source code: anyone can use it, or use pieces of it, to easily make their own products. Each new line of code adds to the total store of knowledge, which helps more development happen faster. At the same time, this can be an open source project's downfall, because it radically lowers the barrier to entry to creating copies. While Web2 companies defend their products from competitors by using proprietary software that would be difficult to copy, open source Web3 projects depend on network effects and reputation over the long term to create moats around their products to defend them from losing users to new market entrants.

An environment where competitors can easily copy the code base of a project means Web3 marketers can't rest on the laurels of building an excellent product with a thriving community. They must not ignore the last step of the marketing funnel: prioritizing strategies to promote stickiness and retain existing users. The value of a Web3 project over time comes not only from the quality of the product and the team behind it but also from its ability to incentivize high-quality contributions from community members, factors that lead it to innovate successful new products and features that a simple copycat wouldn't be capable of creating.

The inherent properties of the technology make the retention game for products vastly different between Web2 and Web3. Whereas the Web2 business model incentivizes building a product and putting a wall around it so no one can see it and copy it, the Web3 model opens up a project's code base for everyone to use, forcing the original project to stay nimble and creative and continue innovating. Ultimately, the former leads to walled gardens and information silos; the latter model incentivizes more high-quality development to happen faster and more collaboratively across an ecosystem. If we could only use one encyclopedia forever and had to choose between Wikipedia and *Encyclopaedia Britannica*, most of us would choose Wikipedia. Because of its large pool of contributors and ability to evolve over time, it's more useful to most of us.

It's logical to expect that the highest-quality products will emerge from projects taking an open source Web3 approach to development. Building in the open has its risks, such as copycats, but these risks are worth it to those who want to build the best products. For Web3 projects, the best moat for defensibility isn't building proprietary software, it's building community and reputation over time.

PART

4

Web3 Transformation and Web2.5

13

Defining Web2.5

WHEN WE STARTED Serotonin, our goal was to use the best practices from our experience at ConsenSys as the first Web3 marketing team—bringing Ethereum to market, along with products such as MetaMask, Infura, and Truffle, as well as developing company brands such as ConsenSys itself—to the next generation of projects building on Web3. That meant DAOs, DeFi protocols, Web3 utilities and infrastructure, Layer-1 and Layer-2 blockchains, as well as NFT projects. Most of our clients were innovative new Web3-native projects referred to us by a small group of dedicated Web3 venture capital firms that knew us by reputation. By 2021, however, NFTs were gaining mainstream adoption, and the hello@serotonin.co inbox was full of outreach from top companies who heard through word-of-mouth that we were the experts on Web3 and reaching the Web3 community. Through the simple contact form on our website, we received emails from the likes of Sony Music, Yum! Brands, and someone pretending to be Kanye West.

With our long memories from years working in the space, we suspected Web3 would feature a heterogeneous mix of Web3-native projects alongside traditional and Web2 brands that were able to adapt their businesses to play in the new environment. Many long-standing contributors to the Web3 movement reflexively reject large brands and corporations entering the space. Remember, this is a community that originally formed in opposition to the institutions that brought about financial crises, the entities extracting giant rents for facilitating transactions, and the Web2 business models handsomely rewarding small numbers of insiders by exploiting vast populations of users and creators.

But anyone who knows the story of the EEA launch should understand the paradoxical relationship between Web3 and the traditional business world. While Web3 offers us a toolkit to push back against outmoded business models, it's equally fueled by the participation of entities who became prominent and powerful under those models. Their endorsements helped the movement take off in the first place; the prices of tokens such as bitcoin and ether have historically been correlated with the performance of the S&P 500 and fluctuated as a result of US Federal Reserve policy; and mainstream institutions adopting Web3 or adding crypto to their balance sheets send positive shock waves through the ecosystem.

At Serotonin, we decided to wade into the paradox, roll up our sleeves, and work with traditional and Web2 companies to help them enter Web3. We started our Web3 transformation practice with the goal of helping these businesses not only use Web3 technologies and reach Web3-native audiences, but also to do so with a natively Web3 approach designed to strengthen Web3 overall. As we stood up this new practice, we realized there was a term missing to describe companies that exist on a spectrum between Web2 and Web3—from a company such as Coinbase that sells customers crypto assets, but with a traditional centralized exchange business model, all the way to a giant retailer such as Nike dipping its toes in Web3 with the launch of CryptoKicks sneaker NFTs—so we did our own exercise in category creation and coined the term *Web2.5* to refer to the projects and companies in between Web2 and Web3. The term caught on across the industry, confirming our suspicion we were onto something.

For existing businesses, the transition from Web2 to Web3 doesn't happen with the flip of a switch. Rather, Web3 transformations progress over time. There is a gradient from Web2 to Web3. Very few projects actually reach the end state of becoming completely decentralized; for many, this is an impossibly high bar. It requires not only decentralizing the code with something like Ethereum but also decentralizing how the application stores data with a tool such as IPFS, and decentralizing governance away from a centralized legal entity into a DAO. Most of the major DEXes and DeFi projects, such as Uniswap and Aave, have their decentralized protocols, but they also have centralized elements such as front-end websites for easier UX or legal entities that exist within the jurisdiction of central governments. Smaller DeFi projects, too, even if decentralization is their main value proposition, often use collateral that is controlled by a centralized company. Frax is an example of a stablecoin project that is highly decentralized at the protocol level but collateralized partly using an asset controlled by a centralized entity—in Frax's case, the token USDC from the payments company Circle, which is known (among other things) to freeze accounts under government pressure.

The Fortune 500 are a long way away from fully decentralizing their business models or technology stacks. Most of the Web2.5 companies we work with in our Web3 transformation practice have never taken a payment in crypto before. From Sotheby's to Pace Gallery, Liverpool Football Club to Roofstock, companies of all sorts approach Serotonin knowing they need a Web3 strategy and looking for partnership. Many, however, simply do not have the problems that full decentralization is intended to solve: the need to resist attack or manipulation by interested parties. Projects really require decentralization only in proportion to having the kind of problem that it solves. At the same time, companies understandably wish to harness the benefits of partial decentralization, or of launching a single, more decentralized project or product line: building economically aligned communities to promote higher engagement and more efficient growth over time, launching valuable membership and rewards programs, leveraging the public blockchain as a powerful CRM, or opening up storefronts in Web3-enabled metaverse worlds. Companies often come to us when their marketing or innovation departments mandate that they have a Web3 strategy or when they identify short-term revenue opportunities in Web3. Once they learn about what is possible with the Web3 substrate, however, many stick around and build their own internal teams to harness these long-term benefits. It is only logical that these companies follow their incentives and pour into Web2.5.

The dream I shared with Joe Lubin during my original ConsenSys interview in 2016—of making Ethereum a household name—has in this sense come true. Often the companies we work with have one or two Web3 enthusiasts leading the charge and trying to convince their cautious CIO or CMO to start investing in their brand's evolution into Web3. We often work with young, passionate leaders of big companies' nascent internal blockchain teams to help them convince their bosses that Web3 presents a unique opportunity. We've seen many of their careers go meteoric after stepping up to lead blockchain. As we support well-known brands spanning a wide variety of industries, our North Star is that brands should enter Web3 the Web3 way. Marketers and other professionals in Web2.5 companies must learn about Web3.

During the crypto bull market of 2021, Twitter was overrun with brands and celebrities hawking their new digital goods. NFTs were exploding in popularity. Jeff Bezos once said he started Amazon, the first online bookstore, when he saw that web use was increasing 2,300% per year; in 2021, NFT trade volume grew a whopping 21,000%.[1] Brands and celebrities hoped to seize the hype to sell NFTs to newly rich crypto kids. Many were surprised to find this a challenge. Web3 natives are intensely skeptical of newcomers trying to sell them anything. If the sellers don't appear authentically interested in Web3, they probably aren't willing to maintain and grow long-term communities, which makes their NFTs more likely to shrink than grow in value over time. The brands and celebrities who were successful in 2021—and continued to grow their communities into 2022 and beyond—are the ones who took this type of long-term approach to their Web3 strategies over trying to make a quick buck. We rigorously apply this insight to our Web3 transformation practice and decline to partner with any organization motivated by only short-term financial interests in their Web3 approach once they understand its deeper benefits.

In this spirit, we offer companies five pillars of a good Web3 strategy. The first pillar is that it strengthens the existing Web3 community, instead of weakening it by diverting attention and resources elsewhere.

The second pillar is that all tokens or NFTs a project launches should have utility—in this context, something community members can do, access, claim, or participate in, that they otherwise wouldn't have. It isn't enough for Nike to launch a Nike fungible token that doesn't do anything or a Nike NFT collection simply featuring pictures of the Nike logo or merchandise. In fact, utility is exactly what Nike focused on when they launched their Nike Cryptokicks NFT sneaker collection, airdropped to the holders of Clone-X NFTs, which Clone-X avatars will be able to actually wear in Web3-enabled metaverse worlds. Nike topped the charts of a 2022 Dune Analytics report analyzing large brands' NFT performance on Ethereum using DEX metrics, earning a whopping $183.82 million in NFT revenue. Dolce & Gabbana, second in line with $23.67 million earned, also focused on utility with their #DGFamily NFT Boxes collection. The ornately carved (virtual) boxes house holders' collections of Dolce & Gabbana digital goods and, as NFTs, grant holders access to a community, rewards, and exclusive sales and events—the different finishes on the boxes indicating different membership tiers. Tiffany & Co., next on the list, programmed utility into their offering with a special twist: for a limited time, they let CryptoPunks NFT holders customize Tiffany jewelry with the NFT they owned. For large brands entering the metaverse, having a well-known name isn't enough. Web3 audiences care about the utility of the digital goods they buy, not just the brand name on the "label."[2]

The third pillar is for brands to take long-term approaches to building communities: staffing projects for indefinite futures rather than limited drops, developing multiyear road maps, and committing teams to deliver against them.

The fourth pillar is for brands to introduce Web3 incentive structures into their business models wherever they touch Web3 so they can build economically aligned communities.

There is a fifth pillar, in our work with retail brands: the launching of metaverse storefronts, activations, and games in real Web3-enabled metaverse worlds such as Decentraland and Sandbox.

At Serotonin, we believe every brand will eventually end up in Web3. When the web was originally created, most of its use cases were military or academic. Few people believed—or even imagined—it would become the primary global destination for commercial and social life. As Web2 pioneered standards and cultivated incentives to come online, that reality emerged, and companies scrambled to start using the internet. Consulting firms cropped up with "digital transformation" practices. Many of these persist today, as traditional companies are still in the process of updating their business models and processes to take advantage of Web2.

Just like digital transformation, we think Web3 transformation will become ubiquitous, as economic and social incentives pull brands, their investors, and customer bases into the decentralized web—even if they do not require full decentralization of every part of their business. If brands start this transformation quickly, they have the opportunity to be heralded as early adopters. Those that lag behind may find that by the time they enter Web3, their niche is already occupied by a Web3-native project. Just as the web killed old businesses and fostered an entire generation of new ones, existing companies are not guaranteed survival in a Web3 world. The best a company can do is move quickly to develop a thoughtful Web3 strategy: one that strengthens the existing Web3 community, takes a long-term view, introduces Web3 incentives, and explores the Web3-enabled virtual worlds that are poised to eventually become the meeting point for so many of our activities.

Sometimes the companies we meet hesitate to start building Web3 strategies, even when there are some passionate Web3 advocates inside the organization. They are typically concerned about one of the following: environmental impact, crypto market cycles, or the reputational risk of publicly championing Web3. Here is how we usually address each of these concerns.

Many companies' target audiences, especially if they are retail businesses, are concerned about climate change. Some have heard and internalized the message that crypto is not environmentally sustainable. Our approach to this concern begins with education. Not all blockchains are created equal, and the proof-of-work mining algorithm used by Bitcoin and previously used by Ethereum has significant energy requirements. However, minting one NFT versus 1 billion NFTs on the Ethereum blockchain, even when it still used proof-of-work, did not change the marginal energy use of the Ethereum network, because it was the activity of mining that used almost 100% of the energy required to power the network, as opposed to generation or transaction of on-chain assets. But today, after the Merge that updated Ethereum to proof-of-stake, the answer to the climate question is easier for everyone to understand: with the transition to PoS, Ethereum has reduced its energy consumption by over 99%, making it an excellent choice of blockchain platform for climate-conscious companies. That being said, until the rollout of sharding, another Ethereum network update that promises to increase scalability, Ethereum requires gas fees for use that can feel substantial to buyers, especially relative to the cost of items if they are buying lower-cost items, or if they are performing many smaller transactions, such as in the context of a game. Depending on a company's Web3 product or target audience, it may make more sense to use a Layer-2 such as Polygon to avoid high gas fees. Some companies will explore alternative Layer-1 blockchains such as Flow and Solana for their Web3 products, but they should be aware that today there is less liquidity on those platforms to put demand pressure on products, fewer users with those blockchains' native wallets already installed, and risk of asset loss on Layer-1 blockchains that have prioritized scalability or other design goals over decentralization. Regardless of platform, companies have several good options for launching Web3 projects in a way that is conscious of climate concerns.

Similarly, there are multiple smart approaches to navigating crypto market cycles. It starts with understanding the target audience for one's Web3 products. Though the price of crypto tokens such as bitcoin and ether, as well as prominent NFT collections such as CryptoPunks and Bored Apes, can be highly volatile, rocketing between low prices in bear markets and reaching new all-time highs in each subsequent bull market, not all buyers of Web3 assets approach their collections like crypto traders. The same way most people don't check the fiat currency markets to see how the dollar is performing against the yuan before buying their morning coffee, a pair of sneakers, or a music album, those buying Web3 assets primarily for utility (using a Web3 rewards program to buy coffee, getting a pair of sneakers for an in-game avatar, buying a music NFT) don't necessarily care about crypto markets; they will look at the price and evaluate whether or not to buy based on the utility of the item and the cost in their local fiat currency.

If a project's target audience is crypto traders, it may make sense to start building the project immediately and then wait to launch it until the optimal moment in the market. The crypto markets have historically been cyclic. Many companies make the mistake of only starting to build projects at the top of a bull market so that by the time they're ready to launch, the market conditions have become unfavorable. Instead, a better approach is to build sooner and be prepared to launch any time, because conditions can shift quickly. Finally, it is often healthy for a project's growth to start small, build a genuine community that cares about a project, and grow alongside the market.

At the bottom of a bear market, it's easier for project teams to get feedback that informs iterations to make their products better. The exuberance of a bull market for Web3 projects often means so many buyers rushing in to purchase tokens that it's hard to get feedback from real users. No matter where we are in a market cycle, Web3 is here to stay, and eventually, just like they needed web strategies, almost every company will need an approach to Web3, so it behooves most to start building against a thoughtful, long-term road map that gives them the optionality to launch at the most strategic future moment.

The trickiest question from companies considering entering Web3, far beyond climate and market cycle concerns, is about reputation. I have lamented throughout this book that a number of scam artists have deceived their communities in order to make a quick buck in Web3. This is similar to the early days of the web, with sham startups that defrauded users and investors sometimes making away with millions of dollars. After the dot-com bubble burst in the late 1990s, web stocks crashed, and the reputation of web companies was badly tarnished. Investors and users became skeptical about any new web project. In 1998, the economist Paul Krugman infamously predicted that "by 2005 or so, it will become clear that the Internet's impact on the economy has been no greater than the fax machine's."[3] A December 2000 headline in *The Daily Mail* read "Internet 'may be just a passing fad as millions give up on it.'"[4] During a crypto bear market, public opinion is just as harsh about Web3, and then, during a bull market when prices once again soar, hackers and scammers come out from hibernation.

But just as companies were clearly, over a sufficient time frame, wrong to be deterred by crashes and scams from adapting to the web, they would be wrong to miss Web3, because it is a similarly paradigm-shifting technology with powerful capabilities that can confer enormous benefits, especially on the companies that adopt it early. Some of the largest companies have already proven up to the challenge. Starbucks and Chipotle aggressively announced Web3 rewards programs in 2022, despite wintry conditions in the crypto markets. Others like Reddit launched Web3 products, but without using terminology such as *NFTs*. Eventually, just like companies today face no reputational hazards when they launch a "web" product, I suspect there will be no backlash when any company announces a product in "Web3," which will become so ubiquitous that it will be unnecessary to use the term.

In the meantime, companies should focus on launching Web3 products that their target audiences actually want and that are priced and distributed appropriately for those audiences. Backlash usually comes when a company or celebrity launches a product that is poorly thought out or an obvious cash grab, as it would for any other product. Some have made mistakes entering Web3 with products that give this appearance, and then confused the resulting backlash for backlash against Web3. Others may learn from their audiences that today they need to avoid blockchain and Web3 language. Either way, there is no excuse for companies to dismiss Web3 as a "passing fad" or "fax machine," and every reason to develop a strategy to harness its benefits in a way that adds value and makes sense from the perspective of their audience.

Companies should start the mission of developing a Web3 strategy by deeply understanding the DNA of what makes them successful with their existing audience in their Web2 or traditional context. In parallel, they should deploy teams and work with experienced partners to learn about the Web3 substrate. Once they understand their own DNA and how Web3 works, they can imagine how to weave the two together into a new symbiotic whole, simultaneously evolving their own brand and innovating in Web3. If this sounds overly poetic, let us focus on the illustrative example of Sotheby's.

Over the centuries following its founding in 1744, Sotheby's developed a reputation for curating the highest-quality and most desirable physical objects and making them available at auction. Thanks to this reputation, it grew a loyal and devoted customer base of collectors and enthusiasts, expanded its footprint globally, and broadened its mandate to sell fashion, real estate, wines, cars, and more. Sotheby's became the destination where the estates of deceased celebrities, real estate on the moon, and real *T. rex* fossils could be bought and sold alongside Picasso, Monet, and Degas. Thanks to its forward-thinking leadership, especially during the pandemic, Sotheby's refocused its business online, growing e-commerce storefronts in addition to its live auctions.

When NFTs began gaining traction in 2021—and especially when rival auction house Christie's sold the famous $69 million Beeple NFT—Sotheby's quickly identified the opportunity. My cofounder Matthew Iles reached out to a friend working at Sotheby's in their technology department, and soon Serotonin and Sotheby's partnered. Working closely together, we helped guide their transition into Web3, which would yield over $100 million in sales in 2021, recognition as an early mover that would earn them a place on the *Time* 100 Most Influential Companies of 2022 list, and a permanent NFT storefront powered by Serotonin's NFT e-commerce software spinout, Mojito.[5] Together we created poetry at the intersection of the abiding, multi-century DNA of Sotheby's and the novel capabilities of Web3.

One of Sotheby's primary value propositions is validating that an object is authentic and desirable. The Sotheby's brand name is the ultimate seal of approval, the third-party confirmation that confers trust in the value of a given auction piece. In a Web3 world with thousands of NFT sellers screaming for attention and potential buyers trying to separate the wheat of long-term value from the chaff, Sotheby's was in a unique position to validate the early idea that scarce digital goods on a decentralized blockchain could hold real value and curate the best art NFTs for its community. The validation from Sotheby's starting to sell NFTs served as a stamp of approval for the entire ecosystem, ushering a deep Rolodex of the wealthiest art patrons and collectors—many of whom had been hesitant about the new medium—into Web3, strengthening the Web3 community.

Sotheby's partnered with existing Web3 projects and influencers, loaning their authority to collections such as CryptoPunks, Bored Apes, and Autoglyphs, and digital artists such as Mad Dog Jones and Pak.[6] It built a long-term NFT storefront, the Sotheby's Metaverse, and fostered a community there that gathers on a Discord server. Sotheby's incorporated crypto payments into their business model, becoming the first auction house to allow payments in ether. In an auction of a copy of the US Constitution, Sotheby's accepted a DAO, ConstitutionDAO, as a bidder. We even collaborated on launching a replica of Sotheby's New York auction house in Decentraland. Checking Decentraland on our phones in the back row of a live auction in New York, the CEO of Sotheby's and I realized there were far more attendees watching the auction from there than in the physical auction house. Occasionally, a bid would come in from the metaverse. It was a futuristic moment, and one Sotheby's can likely expect many times again, now that it is solidly a Web2.5 company.

The Sotheby's case study is proof that no matter how old and storied a brand, it can succeed in Web3 by translating its value proposition into a Web3 context in a way that strengthens the existing Web3 community. Companies entering Web3 and the metaverse can expect resistance from some Web3 native audiences that believe Web3 and the metaverse are no place for Web2 or traditional companies. However, the more new audiences enter Web3, and the more companies prove it's possible to enter Web3 the Web3 way—honoring their promises to their communities and delivering against their road maps over long time horizons—the more acceptable this will become. Eventually, just as commerce and social life migrated from the physical world onto the web, they will move onto Web3, with redesigned, adapted experiences.

Web2 and traditional companies are free to use the tools and mechanisms of Web3. Their first forays into Web3 don't necessarily need to involve selling NFT collections or tokens. They could, for example, be retroactive distributions of claimables rewarding their existing loyal users and fans that enable these audience members to join a token-gated Web3 community. Often, it's a better approach—a show of good faith—to offer something first before asking new users to buy anything. Companies can pair the physical goods they offer with NFTs or use the blockchain CRM to reward holders with a fungible token. They can build storefronts, or their own metaverse worlds—ideally with open, interoperable Web3 standards—and sell or distribute land in them. And they can introduce new features to the buying experience such as gamification, trying on clothing in 3D, or modeling what furniture would look like inside a physical house. Creators and celebrities can offer new mechanisms for fans to engage with them, launching DAOs that hold rights and royalties, weigh in on creative decisions, or vote on aspects of governance.

Web3 is a playground for new and existing projects alike. Over time, the incentives of Web3 should make its denizens welcoming to new entrants, as long as they respect the rules of engagement.

Conclusion

MARKETING IS HAPPENING all the time in Web3, but many of the most successful marketers in the space aren't formally trained and wouldn't identify themselves as marketers. They are simply the people working on Web3 projects who take responsibility for connecting their products with users. Without ever reading a book or taking a course on marketing, they intuit the need to understand one's audience first and foremost, figure out the channels where they live and the problems they care about solving, and craft messaging on those channels to motivate them to overcome their indifference. Over time, these instinctive marketers begin to observe a funnel of users that forms through discovery, leads to engagement, hopefully converts to use, and with enough care and commitment, results in long-term retention. My friend Eva Beylin, who also works in Web3, tweeted something brilliant: "there's either building or trading, everything in between is marketing."[1] In this book I have attempted to codify the "everything in between" in Web3: what works, what doesn't, and how to think about it.

Marketing isn't a limited set of predetermined practices; it's a repeatable process that starts with imagining what it's like to be someone else. In this way, Web3 marketing is no different from any other kind of marketing. What *does* make Web3 marketing different is that much of the marketer's job is intended to be temporary and eventually transfer to the community. Unlike Web2 companies that Meta and Google have contrived to make forever dependent on their platforms, forced to pay to drive revenue and even reach their own audiences, Web3 projects have the opportunity to design self-marketing systems. They need to bootstrap demand and initial community from zero to one using practices such as those described in this book, but once the system is in place, they should make it part of their road map to decentralize the marketing function. Thanks to its potential to decentralize over a network of incentive-aligned community members, Web3 marketing is far more efficient and sustainable than Web2; incentive design and building the right early community are the keys to success. And absolutely anyone working on a Web3 project who is able to imagine what it's like to be a potential user, develop messaging, build a funnel, and measure and optimize against results, is taking the proper approach for a Web3 marketer, regardless of their actual role or title.

Every technology that changes everything appears unlikely at first, but in hindsight, it couldn't have happened any other way. Back in 2016 when I joined ConsenSys, we suspected blockchain-based projects could powerfully incentivize aligned behavior from groups, though at the time there were few examples. Between that moment and today, we've seen the logic of Web3 starting to play out in reality. With the growth of Ethereum and other blockchain networks, the runaway adoption of NFTs, and the traditional and Web2 businesses clamoring to enter the space, Web3 is probably past the point of no return. I expect to see the potential of Web3 and the metaverse fully ramify over the next decade. My conviction that this will happen comes from paying attention to the incentives at play. Companies and creators want to escape the linearly scaling materials economy and build in an abundant world where there is no frontier to close. Users want to invest in their favorite projects, investors want to help project teams build, and builders want to collaborate together with these groups to achieve success as efficiently as possible.

We have started to observe a trend both inside and outside Web3 of average people getting more interested in investing their money in the ways that have traditionally realized the greatest returns for a select few. This interest goes beyond day trading, which already exploded in the early days of the web when technology enabled consumers access to securities markets previously only enjoyed by professionals. Today, millions of retail investors are banging down the doors that have prevented them from accessing the most lucrative opportunities reserved for elite *accredited investors* (in the United States, those with a lawyer or CPA's letter that confirms they own $1 million, excluding the value of their home, or make over $200,000 per year for multiple subsequent years), such as investing in private equity.[2] Platforms like Republic and Robinhood have succeeded by making potentially high-reward opportunities accessible to retail investors. The sudden rise to popularity of groups like the subreddit r/wallstreetbets, a motley collective of retail investors coordinating their moves on Reddit, are demonstrations of this trend, with the power to catapult a stock into the limelight, as it did with GameStop, foiling the strategies of major hedge funds.

Permissionless and accessible to everyone, Web3 is the ultimate example of this trend. Those with or without accreditation can access any crypto token on a DEX or join a DAO to invest collectively with others. People with everyday jobs are starting to identify as investors as well as workers. Over time, more are likely to catch on. I predict that as the future of work evolves, most people will diversify their economic activities to operate simultaneously as creators and investors. With each person taking on multiple blended roles, increasingly the categories of user, builder, and investor will disappear, to be replaced by a single category called *community*.

Web3 projects with incentive-aligned communities are designed to outcompete Web2 companies that only make money in the margins between customer acquisition costs and lifetime value, and rely on intermediaries they don't control for the continuation of their businesses. However, Web3 projects are poised to succeed because they are capable of independent and sustainable growth. They don't depend on third-party platforms for business. They can incentivize growth without paying a piper.

More people worldwide are making it part of their lives to invest in these assets and to join these DAOs and communities. That fact, combined with the inherent advantages of Web3 over Web2 business when it comes to growth, gives Web3 enthusiasts like me confidence that our new model for doing business will change everything. Today this logic is just starting to make sense to a small group of early adopters. Tomorrow when we look back, I suspect it will seem inevitable.

Notes

Preface

1. Jeff Benson, "Ethereum Wallet MetaMask Reports 21 Million Users, Up 420% Since April," *Decrypt*, November 17, 2021, https://decrypt .co/86263/ethereum-wallet-metamask-reports-21-million-users; Elizabeth Howcroft, "NFT Sales Hit $25 Billion in 2021, but Growth Shows Signs of Slowing," Reuters, January 11, 2022, https://www .reuters.com/markets/europe/nft-sales-hit-25-billion-2021-growth-shows-signs-slowing-2022–01–10/.
2. CoinMarketCap, August 30, 2022, https://coinmarketcap.com/.
3. Ryan Daws, "Ethereum Officially Kicks Off Its One Million Devs Initiative," Developer Tech News, January 20, 2020, https://www .developer-tech.com/news/2020/jan/20/ethereum-officially-kicks-its-one-million-devs-initiative/.
4. Josh Stark and Evan Van Ness, "The Year in Ethereum 2021," Josh Stark, Mirror.xyz, January 17, 2022, https://stark.mirror.xyz/q3OnsK7-mvfGtTQ72nfoxLyEV5lfYOqUfJIoKBx7BG1I; Visa, *Annual Report 2021*, https://s29.q4cdn.com/385744025/files/doc_downloads/Visa-Inc_-Fiscal-2021-Annual-Report.pdf.
5. Stark and Van Ness, "The Year in Ethereum 2021"; Visa, *Annual Report 2021*.

6. "Gross Domestic Product (Second Estimate) and Corporate Profits (Preliminary), Second Quarter 2022," U.S. Bureau of Economic Analysis (BEA), August 25, 2022, https://www.bea.gov/news/2022/gross-domestic-product-second-estimate-and-corporate-profits-preliminary-second-quarter; "GDP (Current US$)—World," World Bank Open Data, accessed August 30, 2022, https://data.worldbank.org/indicator/NY.GDP.MKTP.CD.

7. Konstantinos Korakitis, Richard Muir, Simon Jones, and Michael Condon, *State of the Developer Nation, 22nd Edition*, SlashData, April 2022, https://slashdata-website-cms.s3.amazonaws.com/sample_reports/VZtJWxZw5Q9NDSAQ.pdf.

8. International Telecommunication Union Development Sector, *Measuring Digital Development: Facts and Figures 2021*, December 1, 2021, https://www.itu.int/itu-d/reports/statistics/facts-figures-2021/.

9. Bitcoin, the system, is usually capitalized; bitcoin, the token, is not.

Chapter 1: The Evolution of Web1 and Web2

1. Multiple sources in late 1993 attribute the quotation to Gilmore; see "The Net Interprets Censorship as Damage and Routes Around It," Quote Investigator, July 12, 2021, https://quoteinvestigator.com/2021/07/12/censor/.

2. Jim Giles, "Internet Encyclopaedias Go Head to Head," *Nature*, December 14, 2005, https://www.nature.com/articles/438900a.

3. For a helpful summary of issues related to streaming music, see Ben Sisario, "Musicians Say Streaming Doesn't Pay. Can the Industry Change?," *New York Times*, May 7, 2021, https://www.nytimes.com/2021/05/07/arts/music/streaming-music-payments.html. For a compelling argument about digital networks accelerating winner-take-all distributions and hollowing out the middle class, see Jaron Lanier, *Who Owns the Future?* (New York: Simon & Schuster, 2013).

4. Meta Platforms, Inc., *Form 10-K Annual Report for the Fiscal Year Ended December 31, 2021*, U.S. Securities and Exchange Commission (SEC), February 3, 2022, https://www.sec.gov/Archives/edgar/data/1326801/000132680122000018/fb-20211231.htm; Alphabet Inc., *Form 10-K Annual Report for the Fiscal Year Ended December 31, 2021*, U.S. Securities and Exchange Commission (SEC), February 2, 2022, https://www.sec.gov/Archives/edgar/data/1652044/000165204422000019/goog-20211231.htm.

5. R. Fielding, M. Nottingham, and J. Reschke, eds., "RFC 9110: HTTP Semantics," Internet Engineering Task Force, June 2022, https://www.rfc-editor.org/rfc/rfc9110.html.

Chapter 2: The Evolution of Bitcoin and Ethereum

1. Joseph Lubin, "Joseph Lubin: Keynote on Ethereum | Ethereal Summit 2017," YouTube video, 28:20, May 25, 2017, https://youtu.be/72DpZ-hbdVd8. Lubin has published revised excerpts from this speech over multiple essays at LinkedIn; the quoted text appears in Joseph Lubin, "The moment I jumped in. . ." LinkedIn, April 22, 2019, https://www.linkedin.com/pulse/moment-i-jumped-joseph-lubin.
2. Satoshi Nakamoto, "Bitcoin: A Peer-to-Peer Electronic Cash System," October 31, 2008, https://bitcoin.org/bitcoin.pdf.
3. Daniel Phillips, "The Bitcoin Genesis Block: How It All Started," *Decrypt*, February 10, 2021, https://decrypt.co/56934/the-bitcoin-genesis-block-how-it-all-started.

Chapter 3: The Evolution of Ethereum into Web3

1. Ether (ETH), bitcoin (BTC), and other tokens are often referred to by their ticker symbols.
2. Paul Vigna, "Fund Based on Digital Currency Ethereum to Wind Down After Alleged Hack," *Wall Street Journal*, June 17, 2016, https://www.wsj.com/articles/investment-fund-based-on-digital-currency-to-wind-down-after-alleged-hack-1466175033.
3. For one account of virtual goods in early web games, see Matt Mihaly, "The Genesis of the Virtual Goods Model," Virtual Economy Research Network, November 1, 2009, https://virtual-economy.org/the_genesis_of_the_virtual_goo/.
4. A 90% failure rate for startups has become conventional wisdom, though it is often not clear what the definition of "success" is: staying business? Hitting a particular target for returns on investment? Over how much time? The specific percentage of failed startups changes depending on these variables, but it's clear that the odds of success, however quantified, are not in the startup's favor. Some attempts to quantify this include Patrick Ward, "Is It True That 90% of Startups Fail?," NanoGlobals, June 29, 2021, https://

nanoglobals.com/startup-failure-rate-myths-origin/, and Marcus Cook, "The Biggest Misconception on Why Startups Fail," *Inc.*, August 27, 2021, https://www.inc.com/marcus-cook/the-biggest-misconception-on-why-startups-fail.html.

5. Aoyon Ashraf, "Vitalik Buterin Says Ethereum Merge Cut Global Energy Usage by 0.2%, One of Biggest Decarbonization Events Ever," *CoinDesk*, September 15, 2022, https://www.coindesk.com/business/2022/09/15/vitalik-buterin-says-ethereum-merge-cut-global-energy-usage-by-02-one-of-biggest-decarbonization-events-ever/. For an alternate viewpoint on the worldwide energy consumption claim, see Daniel Kuhn, "Did the Ethereum Merge Drop 'Worldwide Electricity Consumption' by 0.2%?," *CoinDesk*, September 19, 2022, https://www.coindesk.com/layer2/2022/09/19/did-the-ethereum-merge-drop-worldwide-electricity-consumption-by-02/.

6. For percentages of physical versus nonphysical money: Jeff Desjardins, "All of the World's Money and Markets in One Visualization," Visual Capitalist, May 27, 2020, https://www.visualcapitalist.com/all-of-the-worlds-money-and-markets-in-one-visualization-2020/.

7. CoinMarketCap, August 30, 2022, https://coinmarketcap.com/; "GDP (Current US$)—World," World Bank Open Data.

8. Lest this sound like I am pro-money laundering: there are all kinds of good and bad reasons people want privacy. We want privacy in the shower. We want privacy at the ATM machine. We'd be uncomfortable in a world of cameras everywhere in the IRL world, even though they would catch/prevent more crimes. Similarly, there are good and bad uses of financial privacy. Money laundering is obviously a bad example. One recent good example: Vitalik Buterin used the Tornado Cash mixer—a protocol that facilitates private transactions recently shut down by the US government for money laundering concerns—to make donations to causes in Ukraine to protect the recipients from hacking and theft attempts. Another good example: companies doing business on-chain don't want to make data public such as the prices they pay for raw materials (in normal business they are well within their rights to do so, but if they're transacting on a public chain, the public would see this). Just because there are some nefarious use cases for financial privacy doesn't mean financial privacy shouldn't exist. And there is so much demand for privacy, especially financial privacy, that DeFi tools that promote privacy are going to exist no matter what.

9. Robert Stevens, "'Engineering Error' Led to $34 Million DeFi Hack, Harvest Finance Says," *Decrypt*, October 27, 2020, https://decrypt .co/46445/engineering-error-34-million-defi-hack-harvest-finance.

10. Chainalysis, *The 2022 Crypto Crime Report*, February 2022, https:// go.chainalysis.com/rs/503-FAP-074/images/Crypto-Crime-Report-2022.pdf. Note that this is just the number for "cryptocurrency theft," not including $2.8 billion in rugpull fraud.

Chapter 4: The Metaverse

1. Sam Shead, "Meta Plans to Take a Nearly 50% Cut on Virtual Asset Sales in Its Metaverse," CNBC, April 13, 2022, https://www.cnbc .com/2022/04/13/meta-plans-to-take-a-nearly-50percent-cut-on-nft-sales-in-its-metaverse.html.

2. Max Roser, Hannah Ritchie, and Esteban Ortiz-Ospina, "Internet," Our World in Data, 2015, https://ourworldindata.org/internet.

3. Meghan Bobrowsky, "Facebook Feels $10 Billion Sting from Apple's Privacy Push," *Wall Street Journal*, February 3, 2022, https://www.wsj .com/articles/facebook-feels-10-billion-sting-from-apples-privacy-push-11643898139.

Chapter 5: Inventing Web3 Marketing

1. Rob Urban and David M. Levitt, "Crypto Pioneers Head to Brooklyn to Reshape Finance," Bloomberg, May 7, 2018, https://www .bloomberg.com/news/articles/2018–05–07/-cryptolandia-blockchain-pioneers-take-root-in-hipster-brooklyn#xj4y7vzkg.

2. Nathaniel Popper, "Business Giants to Announce Creation of a Computing System Based on Ethereum," *New York Times*, February 27, 2017, https://www.nytimes.com/2017/02/27/business/dealbook/ ethereum-alliance-business-banking-security.html; Paul Vigna, "The Newest Bank Blockchain: Will This Be the Breakthrough?," *Wall Street Journal*, February 27, 2017, https://www.wsj.com/articles/the-newest-bank-blockchain-will-this-be-the-breakthrough-1488285211; Olga Kharif, "Microsoft, Others Said to Form Group Backing Ethereum Blockchain," Bloomberg, February 9, 2017, https://www.bloomberg .com/news/articles/2017–02–09/microsoft-others-said-to-form-group-backing-ethereum-blockchain; Anna Irrera, "JPMorgan, Microsoft,

Intel and others form new blockchain alliance," Reuters, February 27, 2017, https://www.reuters.com/article/us-ethereum-enterprises-consortium-idUSKBN1662K7.

3. Jim Collins memorably uses the concept of a flywheel to describe the compounding, momentum-generating effect of good effort and decisions in *Good to Great* (New York: HarperBusiness, 2001) and *Turning the Flywheel: A Monograph to Accompany Good to Great* (New York: HarperCollins, 2019).

Chapter 6: Know Your product and Your Audience

1. This virtuous cycle is the classic method for product development from Eric Ries's canonical classic, *The Lean Startup* (New York: Crown Business, 2011).

2. Elizabeth Lopatto, "NBA Top Shot seemed like a slam dunk—so why are some collectors crying foul?," The Verge, June 7, 2022, https://www.theverge.com/23153620/nba-top-shot-nft-bored-ape-yacht-club.

3. Andrew R. Chow, "*Time*100 Most Influential Companies of 2022: Sotheby's," *Time*, March 30, 2022, https://time.com/collection/time100-companies-2022/6159438/sothebys/.

Chapter 8: Set Key Metrics

1. Our standard reference for OKR is Christina Wodtke, *Radical Focus: Achieving Your Most Important Goals with Objectives and Key Results* (Oklahoma City: Cucina Media, 2016).

2. Nick Bostrom explains his paperclip maximizer AI thought experiment: "Suppose we have an AI whose only goal is to make as many paper clips as possible. The AI will realize quickly that it would be much better if there were no humans because humans might decide to switch it off. Because if humans do so, there would be fewer paper clips. Also, human bodies contain a lot of atoms that could be made into paper clips. The future that the AI would be trying to gear towards would be one in which there were a lot of paper clips but no humans." In Kathleen Miles, "Artificial Intelligence May Doom the Human Race Within s Century, Oxford Professor Says," *HuffPost*, August 22, 2014, https://www.huffpost.com/entry/artificial-intelligence-oxford_n_5689858. Original articulation in Nick Bostrom, "Ethical Issues in Advanced Artificial Intelligence," in *Cognitive, Emotive and Ethical Aspects of*

Decision Making in Humans and in Artificial Intelligence, Vol. 2, ed. Iva Smit et al. (Windsor, Ontario: Int. Institute of Advanced Studies in Systems Research and Cybernetics, 2003), 12–17, available in "slightly revised" form at https://nickbostrom.com/ethics/ai.

Chapter 10: Discovery: Break Through

1. Charles Duhigg, "The Craving Brain: How to Create New Habits," in *The Power of Habit* (New York: Random House, 2012). For Hopkins's own account, see Claude C. Hopkins, "Pepsodent," in *My Life in Advertising* (New York: Harper & Brothers, 1927), 151–156, https://books.google.com/books?id=jvhCAAAAIAAJ. There is evidence that the success of Hopkins's advertising led to the incorrect overturning of scientific consensus that oral hygiene and "clean" teeth are cosmetic issues that have nothing to do with actual tooth decay. See Philippe P. Hujoel, "Historical Perspectives on Advertising and the Meme That Personal Oral Hygiene Prevents Dental Caries," *Gerodontology* 36 (2019): 36–44, https://doi.org/10.1111/ger.12374.

2. Camila Russo, "Sale of the Century: The Inside Story of Ethereum's 2014 Premine," *CoinDesk*, July 11, 2020, https://www.coindesk.com/markets/2020/07/11/sale-of-the-century-the-inside-story-of-ethereums-2014-premine/.

3. Jillian D'Onfro, "Google Ends Cryptocurrency Ad Ban—but Only for Certain Kinds of Ads," CNBC, September 25, 2018, https://www.cnbc.com/2018/09/25/google-reverses-ban-on-cryptocurrency-exchange-advertising-in-us-japan.html.

4. Matthew Leising, "Jay-Z, Charles Schwab-Backed Ethereum App Opens Doors to Public," Bloomberg, August 11, 2020, https://www.bloomberg.com/news/articles/2020–08–11/jay-z-charles-schwab-backed-ethereum-app-opens-doors-to-public.

5. Matthew Leising, "Crypto Push by Republic Platform Sparked by New Token," Bloomberg, June 24, 2020, https://www.bloomberg.com/news/articles/2020–06–24/crypto-push-by-republic-investment-platform-sparked-by-new-token.

6. Rob Price, "Weed, Times Square, and Floyd Mayweather: How Cryptocurrency Mania Is Creeping into the Mainstream," *Business Insider*, September 1, 2017, https://www.businessinsider.com/cryptocurrency-bitcoin-ethereum-ico-mania-going-mainstream-2017-8.

7. Angus Berwick and Elizabeth Howcroft, "From Crypto to Christie's," Reuters, November 17, 2021, https://www.reuters.com/investigates/special-report/finance-crypto-sundaresan/.

Chapter 12: Retention: Build a Moat

1. Kati Chitrakorn, "Prada Teams Up with Adidas to Launch First NFT: Hint, Its Beeple-style," Vogue Business, January 20, 2022, https://www.voguebusiness.com/technology/prada-teams-up-with-adidas-to-launch-first-re-source-nft.
2. "Join FWB," Friends With Benefits, accessed August 28, 2022, https://www.fwb.help/join.
3. "Mindmap," Azuki, accessed August 28, 2022, https://www.azuki.com/mindmap.
4. Eli Tan, "Azuki NFT Founder Admits to Abandoning Past Projects," CoinDesk, May 9, 2022, https://www.coindesk.com/business/2022/05/10/azuki-nft-founder-admits-to-abandoning-past-projects/; Stefan Stankovic, "Azuki NFT Review: The Anime Avatar Project Killed by Its Founder," Crypto Briefing, August 25, 2022, https://cryptobriefing.com/azuki-nft-review-the-anime-avatar-project-killed-by-its-founder/.

Chapter 13: Defining Web2.5

1. I first saw this comparison in a tweet by Andrew Steinwold (@AndrewSteinwold, April 7, 2022). For the original figures, see Jeff Bezos interviewed by Richard Wiggins, "The 'lost' Jeff Bezos interview just about a year after starting Amazon," taped June 1997, YouTube video, 6:31, https://youtu.be/Pgzi_jUBu9U, and Ryan Browne, "Trading in NFTs Spiked 21,000% to More Than $17 Billion in 2021, Report Says," CNBC, March 10, 2022, https://www.cnbc.com/2022/03/10/trading-in-nfts-spiked-21000percent-to-top-17-billion-in-2021-report.html.
2. Noah Levine (kingjames23), "NFT Brands Case Study Overview," Dune Analytics, August 11, 2022, https://dune.com/kingjames23/nft-project-possible-data-to-use; Maghan McDowell, "Nike and Rtfkt Take on Digital Fashion with First 'Cryptokick' Sneaker," Vogue

Business, April 23, 2022, https://www.voguebusiness.com/technology/nike-and-rtfkt-take-on-digital-fashion-with-first-cryptokick-sneaker.

3. Paul Krugman, "Why Most Economists' Predictions Are Wrong," *Red Herring*, June 10, 1998, https://web.archive.org/web/19980610100009/https://www.redherring.com/mag/issue55/economics.html.

4. James Chapman, "Internet 'May Be Just a Passing Fad as Millions Give Up on It,'" *The Daily Mail*, December 5, 2000, 33. Images of this article have been extensively posted (and mocked) on Twitter, Reddit, and elsewhere.

5. Crystal Kim, "Sotheby's Makes $100 Million in NFT Sales with Younger Audience," Bloomberg, December 15, 2021, https://www.bloomberg.com/news/articles/2021–12–15/sotheby-s-makes-100-million-in-nft-sales-with-younger-audience.

6. Kim, "Sotheby's"; "Autoglyph #177," Sotheby's, Natively Digital: A Curated NFT Sale, auction date June 10, 2021, https://www.sothebys.com/en/buy/auction/2021/natively-digital-a-curated-nft-sale-2/autoglyph-177; 'Mad Dog Jones, Visor," Sotheby's, Natively Digital: A Curated NFT Sale, auction date June 10, 2021, https://www.sothebys.com/en/buy/auction/2021/natively-digital-a-curated-nft-sale-2/visor; "Mad Dog Jones, Nightfall," Sotheby's, Contemporary Discoveries, auction date July 19, 2022, https://www.sothebys.com/en/buy/auction/2022/contemporary-discoveries-4/nightfall. The biggest Mad Dog Jones NFT sale ($4.1 million) was actually not with Sotheby's, but with Phillips: Sam Gaskin, "Mad Dog Jones' Fax Machine NFT Sells for $4.1m," Ocula, April 23, 2021, https://ocula.com/magazine/art-news/mad-dog-jones-fax-machine-nft-fetches-millions/.

Conclusion

1. Eva Beylin (@evabeylin), Twitter, November 18, 2020, https://twitter.com/evabeylin/status/1329108658941136896.

2. SEC Office of Investor Education and Advocacy, "Accredited Investors – Updated Investor Bulletin," Investor.gov, April 14, 2021, https://www.investor.gov/introduction-investing/general-resources/news-alerts/alerts-bulletins/investor-bulletins/updated-3.

Acknowledgments

I AM GRATEFUL to my husband, Sam Cassatt, for his patience and support throughout the process of writing this book, even when it took place in a seaside town during our summer vacation. A computer scientist by training, Sam was invaluable as a technical reviewer. As a lover of truth, he inspired me to be as precise as possible. Sam was the first person I met in the Ethereum community. It's thanks to him that I discovered this movement.

Next, I want to acknowledge Joe Lubin, who took a chance on me professionally early in my career, offering me the opportunity of a lifetime to do something I had never done before. His belief in me throughout our time working together gave me the confidence to be creative and take risks.

Building the first Web3 marketing team, first at ConsenSys and then at Serotonin, was a group effort. Without the singular genius of my teammates Elise Ransom, Kara Miley, and Everett Muzzy, Web3 would not be where it is today. Nearly seven years in, every day I feel honored to work with you. Thank you also to AJ Banon for your trust and partnership growing the Serotonin business.

Last, I would like to express my gratitude to the team involved directly with this book. Thank you to the editors at Wiley, especially Victoria Savanh, for advocating for this project. Your enthusiasm is what finally pushed me over the edge to write it. And thank you to Matt Morello for your assistance with research, citations, and copyediting. I could not have done this without you—certainly not on time.

About the Author

Amanda Cassatt (née Gutterman) is a notable Web3 product builder and master marketer who has brought some of the leading blockchain companies to market and advises global institutions on how to adopt crypto.

Cassatt is the founder and CEO of Serotonin, the leading Web3 marketing agency and product studio. She is a cofounder and the president of Mojito, the commerce suite for selling digital goods in Web3 that is used by Sotheby's, CAA, and other top brands—the first software company to spin out of Serotonin, in 2021. She additionally advises and serves on the board of numerous other Web3 startups.

From 2016 to 2019, Cassatt brought Ethereum to market as the CMO of ConsenSys, supporting its growth to become the largest and most robust blockchain ecosystem. Prior to joining ConsenSys, she cofounded media startup Slant, which she launched after working under Arianna Huffington at the Huffington Post, where she learned about the challenges for creators in Web2 that she would spend her subsequent career addressing.

Cassatt was named as part of the *Forbes* 30 Under 30 class of 2016 at the age of 24. Having introduced Ethereum and Web3 on stages from TechCrunch Disrupt to the World Economic Forum and SXSW, she regularly speaks at events around the world. She can be found on Twitter at @amandacassatt, where she goes by amanda.eth.

Index

211